SWITCH MODE POWER SUPPLY APPLICATIONS

JOE MARRERO and **RAYMOND ZHANG**

author_HOUSE_

AuthorHouse™
1663 Liberty Drive
Bloomington, IN 47403
www.authorhouse.com
Phone: 1 (800) 839-8640

Published by AuthorHouse 06/22/2020

ISBN: 978-1-7283-4895-7 (sc)
ISBN: 978-1-7283-4896-4 (e)

Library of Congress Control Number: 2020904078

Print information available on the last page.

This book is printed on acid-free paper.

Contents

Nomenclature Symbol defined in the Book

G	Generator Voltage, convention used by Middlebrook
G_{od}	Generator's output to Duty Cycle transfer function
G_{og}	Output to generator voltage transfer function
GV_d	Defined as Generator's Output due to Duty Cycle
V	Defined as Output due to Duty Cycle
V_g	DC Voltage Generator
ω_o	Radians of LC Network
R_o	Characteristic Impedance of LC Network
f_o	Characteristic frequency of LC Network
Q_c	Que that the Capacitor sees
Q_L	Que that the Inductor sees
Q	Overall Que
z & p	Zeroes and Poles
V_{in}	DC Voltage Generator or Input Voltage Source
V_m	Defined as Ramp Height
H(s)	Design of LCR Filter Network
A(s)	Design of Error Amplifier Transfer Function
T	Refers to ΔOpen Loop Gain
$\dfrac{T}{1+T}$	Refers to Close Loop Gain
D	Refers to Duty Cycle at "ON" time
D'	Refers to Duty Cycle at "OFF" time
D+D'	D+D' =1 refers to CCM 100% of switching time
D_2	Refers to Discontinuous Mode
f_{sw}	Switching frequency
f_{BW}, f_μ	Refers to frequency Bandwidth at unity Gain or 0dB
T_s	Refers to switching period $T_s = 1/f_{sw}$
$\hat{V}$	Small signal output voltage
$\hat{\imath}$	Small signal inductor current
$\hat{d}$	Small signal duty cycle variation
$\hat{V}_g$	Small signal input voltage
V_D, V_d	Diode voltage drop
V_o, V_{out}	Output DC voltage
V_{in}	Input DC voltage
ESR	R_{ESR} refers to Capacitor's Equivalent Series Resistance
R_w	Refers to Inductor's DC or wire resistance
R_{dson}	Refer to MOSFET's "ON" resistance
R_d	Refers to switching Diode's parasitic resistance
P	Total Power

P_{in}	Input Power
P_{out}	Output Power
P_{loss}	Power Loss
I_{avg}	Average Current
I_{rms}	Root Mean Square Current
I_{peak}, I_{pk}	Peak Current
(s)	S, Laplace or Frequency domain
(t)	time domain
ΔI	peak to peak current
Z_o, Z_{out}	Output Impedance
Z_{OL}	Output Open Loop Impedance
Z_{CL}	Output Close Loop Impedance
$\hat{V}_Y$	AC Signal Injection vs. GND at Output
$\hat{V}_X$	AC Signal Injection vs. GND at Output
ΔV	Delta Voltage Difference
$A(0)$	DC Gain
A	Gain of Operational Amplifier
BW	Bandwidth
V_{ref}	Voltage Reference
$\hat{V}_{ref}$	Small signal injection at Voltage Reference
s	$s = j\omega$, s is Laplace's frequency domain

DC Switching Regulator Characteristics

with and without Parasitics

Introduction to Switching Power Supply:

Basic Switching Regulator Topologies:

There are three basic power stages which are defined as Buck, Boost and Buck-Boost converter stages. All other topologies are derived from these basic stages. The most common one used in industrial and aerospace applications is the Buck Converter stage.

Buck stage transfers the electrical energy from the input to the output during the conduction or "ON" time of the switch. The output voltage is less than the input voltage and has the same polarity as the input voltage unless a transformer is used between input and output.

The Boost stage transfers the electrical energy from the input to the output during the "OFF" time of the switch. The output voltage is greater than the input voltage and has the same polarity unless a transformer is used. During "ON" time, energy is stored in the inductor. When the switch is turned "OFF", the source energy stored in the inductor is transformed to the load.

The Buck-Boost stage is also called Flyback converter. The electrical energy is stored in the inductor during the "ON" time of the switch, and during the "OFF" time the energy is transferred to the load. The output can be greater or less than the input voltage; and the polarity is reversed unless a transformer is used.

We will derive and explore the DC input/output transfer functions of the three non-isolated power converter stages in this book.

1. BUCK CONVERTER

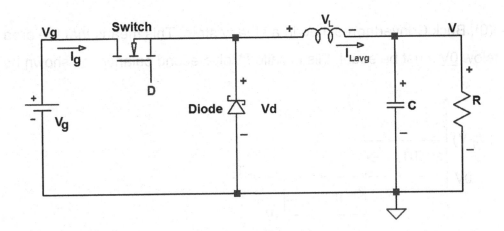

Look at the Voltage across the inductor and the current through the inductor:

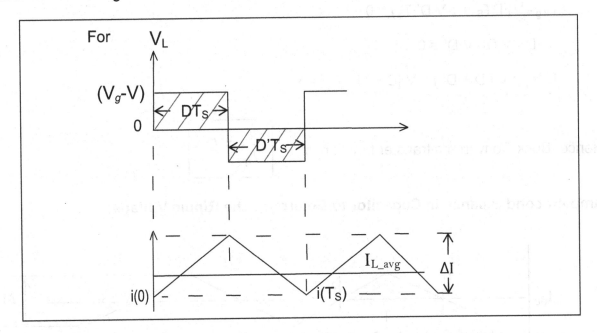

Note: $I_{L\_avg} = I_{dc}$ goes to the load to create the DC output voltage V, and the ripple current, ΔI, goes into the capacitor.

$$V = I_{dc} \cdot R$$

Buck Converter Volt-Second Balance:

Volt-Second balance is nothing but Faradays Law in steady state.

If $i(T_s) = i(0)$, Buck Converter reached in a steady state. This means that the area above 0V and below 0V must be equal, this is called Volt-Second balance, as shown below.

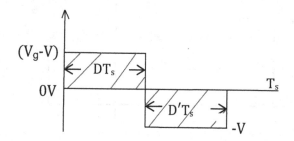

$(V_g - V) \cdot DT_S + (-V \cdot D' \cdot T_S) = 0$

$V_g \cdot D - V \cdot D - V \cdot D' = 0$

$D \cdot V_g = V \cdot [D + D'] = V \cdot [D + (1 - D)] = V$

Hence, Buck Converter's transfer function is:

$$\frac{V}{V_g} = D$$

Amp–Second Balance in Capacitor to Determine the Ripple Voltage:

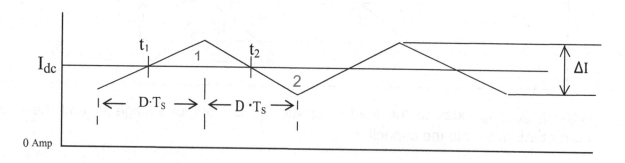

Area 1 must equal to area_2.

If area_1 > area_2, the capacitor voltage will increase.

If area_2 > area_1, the capacitor voltage will decrease.

Let's look at area_1 and note that:

$$C \frac{dv}{dt} = i$$

$$\int_{0}^{\Delta v} dv = \frac{1}{c} \cdot \left(\int_{t_1}^{t_2} i \cdot dt \right) = \frac{T_s}{c} \cdot (\text{area\_1})$$

Since area_1 $\triangleq \left(\frac{1}{T_s} \cdot \int_{t_1}^{t_2} i \cdot dt \right)$

Area_1 $= \frac{1}{Ts} \cdot \frac{1}{2} \left(\frac{D \cdot T_s}{2} + \frac{D' \cdot T_s}{2} \right) \cdot \frac{\Delta I}{2}$, where (D + D') = 1

Area_1 $= \frac{\Delta I}{8}$

Capacitor Ripple $\longrightarrow$ $\boxed{\Delta V = \frac{T_s}{C} \cdot \frac{\Delta I}{8} = \frac{\Delta I}{8 \cdot C \cdot f_s}}$

If we have ESR, then:

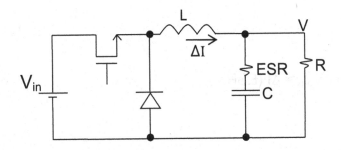

NOTE: at high frequency we have $X_c = \dfrac{1}{2\pi \cdot f \cdot c} \ll \text{ESR}$

$\Delta V = \Delta I(\text{ESR} \| R) \approx \Delta I \cdot \text{ESR}$, where ESR << R

Hence if ESR << R, then all of ΔI or ripple current will go to ESR and capacitor, and average DC current I_{DC} will go to the load R.

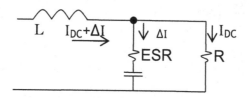

The ripple is a combination of the ESR and the output capacitor C. Since they are out of phase, we must add them vectorially.

$$\Delta V_{p\text{-}p} = \sqrt{\Delta V_{ESR}^2 + \Delta V_{cap}^2}$$

$$\Delta V_{ESR} = \Delta I \cdot (ESR)$$

$$\Delta V_{cap} = \frac{\Delta I}{8 \cdot c \cdot f_s}$$

Generally $\Delta V_{ESR} > \Delta V_{cap}$, then $\Delta V_{p\text{-}p} \approx \Delta I \cdot (ESR)$

1.1 Buck Converter with Parasitics

Voltage−Second Balance:

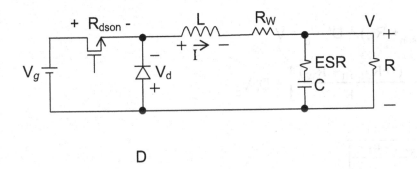

$$D \qquad\qquad\qquad\qquad\qquad\qquad\qquad D'$$

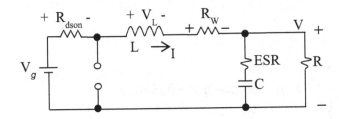

$$L\frac{di}{dt} = V_L = (V_g - I \cdot R_{dson} - I \cdot R_W - V)$$

$$L\frac{di}{dt} = V_L = (-V_d - I \cdot R_W - V)$$

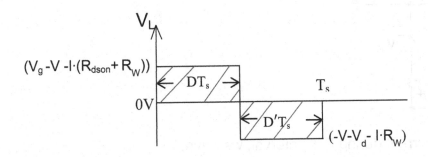

Areas must equal for steady state (Volt-Second balance).

Therefore

$$D \cdot (V_g - I \cdot R_{dson} - I \cdot R_W - V) + D'(-V_d - I \cdot R_W - V) = 0$$

$$D \cdot V_g - I \cdot D \cdot R_{dson} - I \cdot D \cdot R_W - V \cdot D - D' \cdot V_d - I \cdot D' \cdot R_W - V \cdot D' = 0$$

$$D \cdot V_g - I \cdot D \cdot R_{dson} = D' \cdot V_d + I \cdot R_W \cdot (D + D') + V(D + D') = D' \cdot V_d + I \cdot R_W + V$$

Amp–Second Balance:

For steady state, $I = I_{dc}$ must all flow to the load, R.

Hence $I = V/R$

$$\therefore \quad D \cdot V_g = I \cdot (D \cdot R_{dson} + R_W) + D' \cdot V_d + V$$

$$D \cdot V_g = V\left[1 + \frac{(D \cdot R_{dson} + R_w)}{R}\right] + D' \cdot V_d$$

$$V = \frac{(D \cdot V_g - D' \cdot V_d)}{\left[1 + \frac{(D \cdot R_{dson} + R_w)}{R}\right]}$$

$$\frac{V}{V_g} = \frac{(D - D' \cdot V_d / V_g)}{\left[1 + \frac{(D \cdot R_{dson} + R_w)}{R}\right]}$$

$$\frac{V}{V_g} = D \cdot \left\{ \frac{\left(1 - \left(\frac{D'}{D}\right)\left(\frac{V_d}{V_g}\right)\right)}{\left[1 + \frac{(D \cdot R_{dson} + R_w)}{R}\right]} \right\} = D \cdot \eta, \quad \text{where } \eta = \text{Efficiency}$$

Note: By inspection we can add a series resistance, R_d, to the diode.

Diode mode:

Since R_d only see's current during $D'T_S$ interval, we have:

$$\boxed{\frac{V}{V_g} = D \left\{ \frac{\left(1 - \left(\frac{D'}{D}\right)\left(\frac{V_d}{V_g}\right)\right)}{1 + \frac{(D \cdot R_{dson} + D' R_d + R_w)}{R}} \right\} = D \cdot \eta}$$

1.2 Buck Converter Efficiency & Gain Plots vs. R$_{dson}$ in MOSFET

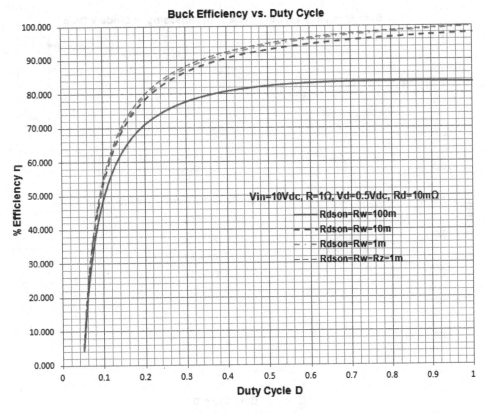

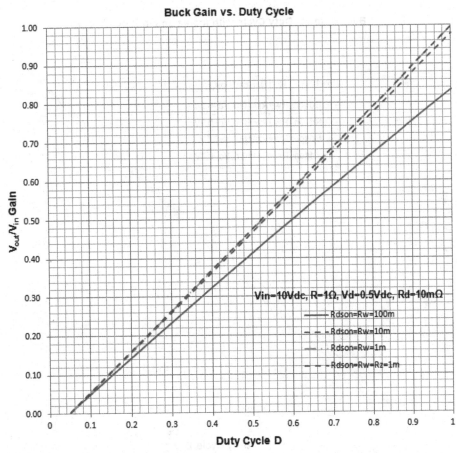

1.3 Buck Converter Efficiency & Gain Plots vs. Diode Voltage Drop

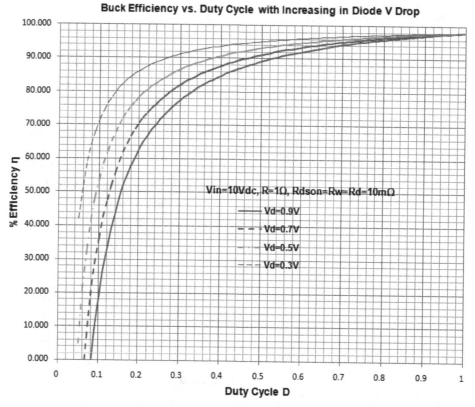

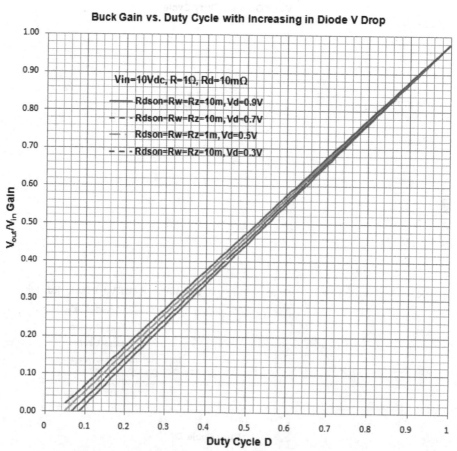

1.4 Buck Converter Efficiency & Gain Plots with Synchronous Rectifier

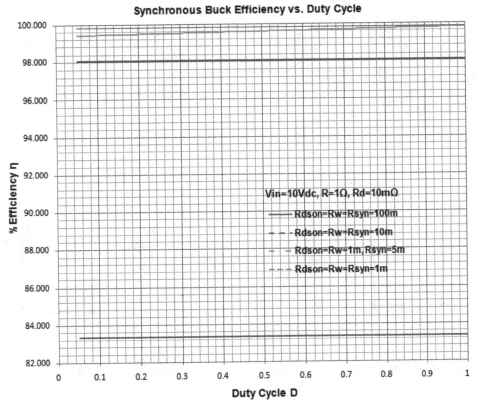

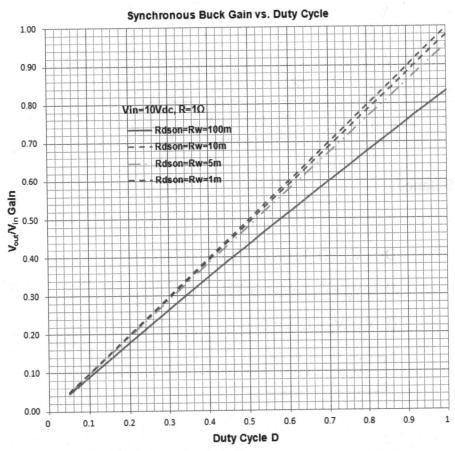

2. BOOST CONVERTER

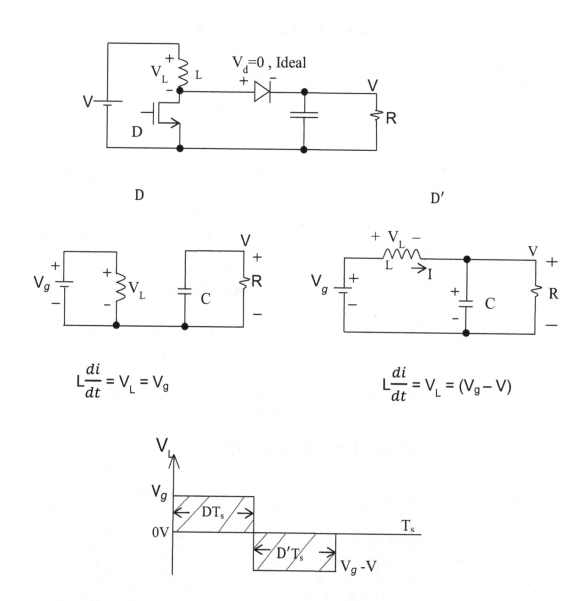

Volt – Sec Balance:

$$D \cdot T_s \cdot V_g + D' \cdot T_s \cdot (V_g - V) = 0$$

$$D \cdot T_s \cdot V_g + D' \cdot T_s \cdot V_g - D' \cdot T_s \cdot V = 0$$

$$(D + D')V_g - D' \cdot V = 0$$

$$\boxed{V/V_g \quad 1/D'}$$

2.1 Boost Converter with Parasitics

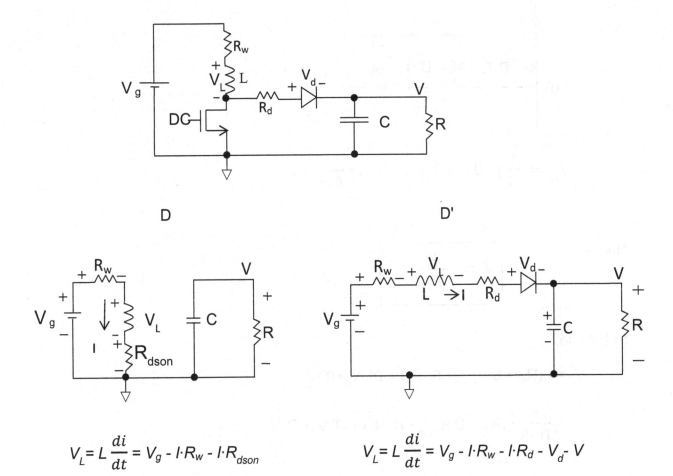

$$V_L = L \frac{di}{dt} = V_g - I \cdot R_w - I \cdot R_{dson}$$

$$V_L = L \frac{di}{dt} = V_g - I \cdot R_w - I \cdot R_d - V_d - V$$

Volt – Sec Balance:

$$D \cdot V_g - I \cdot D \cdot R_w - I \cdot D \cdot R_{dson} + D'V_g - I \cdot D' \cdot R_w - I \cdot D' \cdot R_d - D' \cdot V_d - D' \cdot V = 0$$

$$V_g - I \cdot R_w - I \cdot D \cdot R_{dson} - I \cdot D' \cdot R_d - D' \cdot V_d - D' \cdot V = 0$$

Amp – Sec Balance:

I_{avg} must flow through the load (R) or we will not be in steady state.

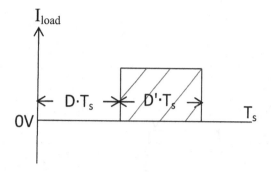

$$I_{avg} = \frac{1}{T_s}(I \cdot D' \cdot T_s) = I \cdot D' = \frac{V}{R}$$

Hence:

$$\boxed{I = \frac{V}{D' \cdot R}}$$

Substitute:

$$V_g = I \cdot (R_w + D \cdot R_{dson} + D' \cdot R_d) + D' \cdot V_d + D' \cdot V$$

$$= \frac{V}{D' \cdot R} \cdot (R_w + D \cdot R_{dson} + D' \cdot R_d) + D' \cdot V_d + D' \cdot V$$

Move $D' \cdot V_d$ to the left, and pull V out on the right side of equation:

$$V_g - D'V_d = V \cdot \left[D' + \frac{(R_w + D \cdot R_{dson} + D' \cdot R_d)}{D' \cdot R} \right]$$

Pull D' out from right side of equation, we have

$$1 - D' \cdot \frac{V_d}{V_g} = \frac{V}{V_g} \cdot D' \left[1 + \frac{(R_w + D \cdot R_{dson} + D' \cdot R_d)}{D'^2 \cdot R} \right]$$

$$\boxed{\frac{V}{V_g} = \frac{1}{D'} \left\{ \frac{\left(1 - D'\left(\frac{V_d}{V_g}\right) \right)}{\left[1 + \frac{(R_w + D \cdot R_{dson} + D' \cdot R_d)}{D'^2 \cdot R} \right]} \right\} = \frac{1}{D'} \cdot \eta}$$

2.2 Boost Converter Efficiency & Gain Plots vs. R_{dson} in MOSFET

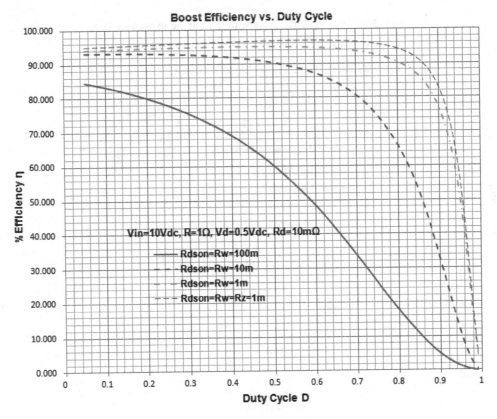

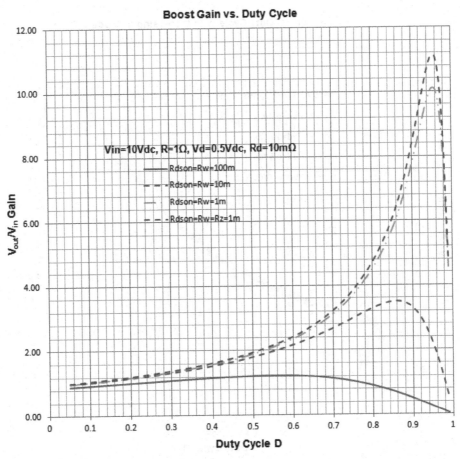

2.3 Boost Converter Efficiency & Gain Plots vs. Diode Voltage Drop

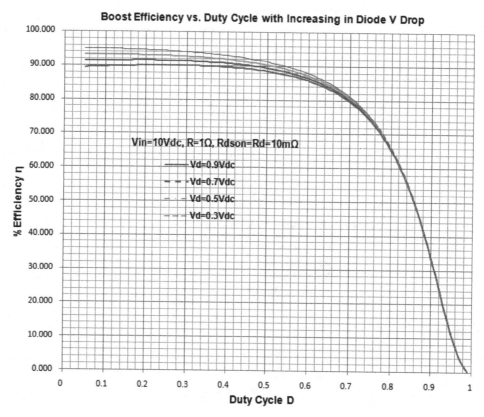

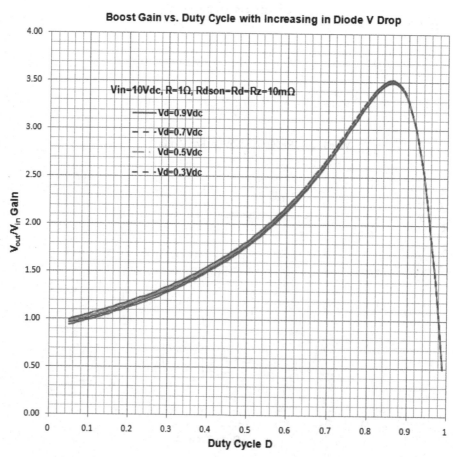

2.4 Boost Converter Efficiency & Gain Plots with Synchronous Rectifier

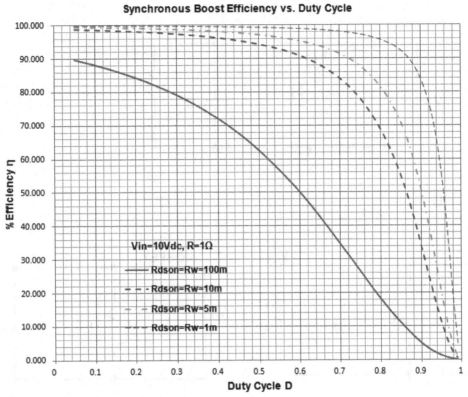

Synchronous Boost Efficiency vs. Duty Cycle

Vin=10Vdc, R=1Ω

—— Rdson=Rw=100m

— — Rdson=Rw=10m

— · — Rdson=Rw=5m

— — — Rdson=Rw=1m

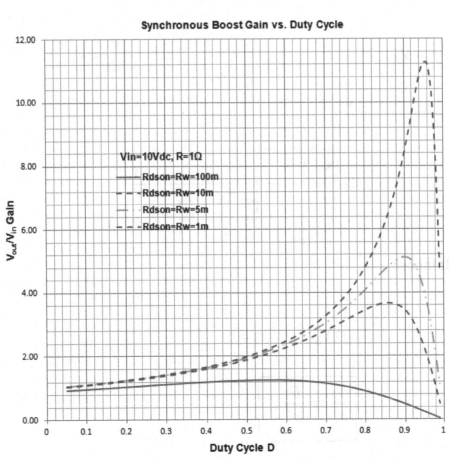

Synchronous Boost Gain vs. Duty Cycle

Vin=10Vdc, R=1Ω

—— Rdson=Rw=100m

— — Rdson=Rw=10m

Rdson=Rw=5m

— — — Rdson=Rw=1m

3. BUCK-BOOST CONVERTER

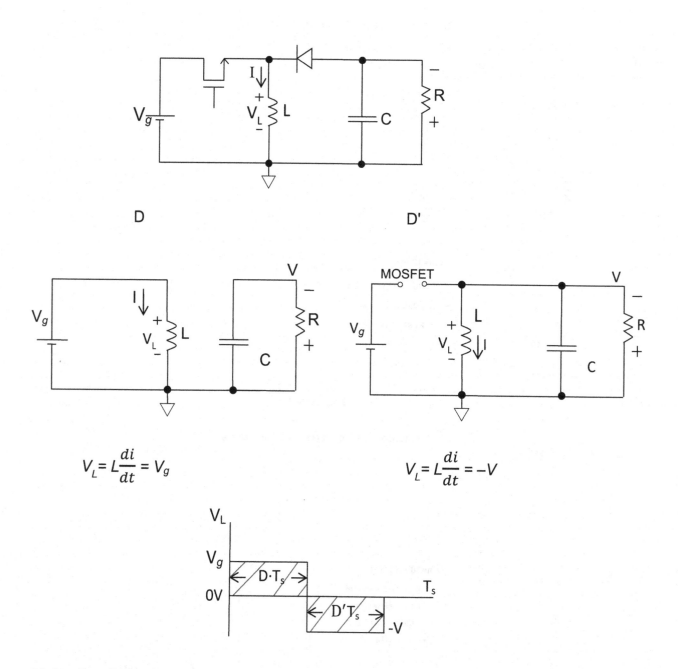

$$V_L = L\frac{di}{dt} = V_g$$

$$V_L = L\frac{di}{dt} = -V$$

Volt - Sec Balance:

$$D\cdot V_g + D'\cdot(-V) = 0$$

$$\boxed{\frac{V}{V_g} = \frac{D}{D'}}$$

3.1 Buck-Boost Converter with Parasitics:

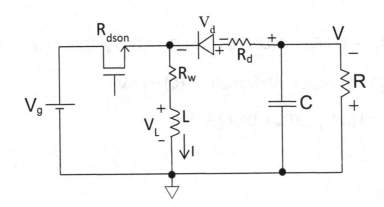

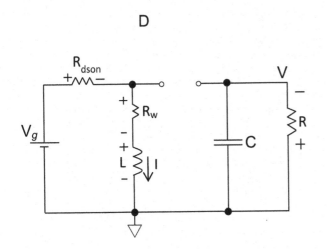

D

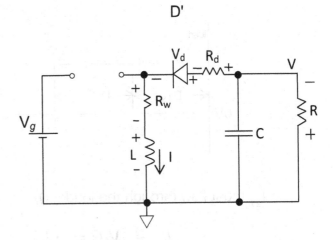

D'

$$V_L = L\frac{di}{dt} = V_g - I \cdot R_{dson} - I \cdot R_w$$

$$V_L = L\frac{di}{dt} = -I \cdot R_w - I \cdot R_d - V_d - V$$

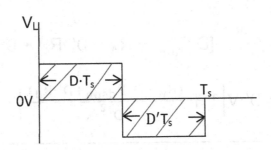

Volt – Sec Balance:

$$D \cdot V_g - D \cdot I \cdot R_{dson} - D \cdot I \cdot R_W + (-D' \cdot I \cdot R_w - D' \cdot I \cdot R_d - D' \cdot V_d - D' \cdot V) = 0$$

$$D \cdot V_g - D' \cdot V_d = I \cdot [D \cdot R_{dson} + (D+D')R_w + D' \cdot R_d] + D' \cdot V$$

$$D \cdot V_g - D' \cdot V_d = I \cdot [D \cdot R_{dson} + R_w + D' \cdot R_d] + D' \cdot V$$

Amp – Sec Balance:

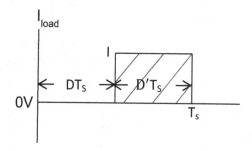

I_{avg} must flow through the load (R),

$$I_{avg} \triangleq V/R = I \cdot D'$$

$$\therefore \quad \boxed{I = \frac{1}{D'} \cdot \frac{V}{R}}$$

<u>Substituting</u>

$$D \cdot V_g - D' \cdot V_d = \frac{V}{D' \cdot R} [D \cdot R_{dson} + R_W + D' \cdot R_d] + D' \cdot V$$

$$D(V_g - \frac{D'}{D} \cdot V_d) = D' \cdot V \left[1 + \frac{(R_w + D \cdot R_{dson} + D' \cdot R_d)}{D'^2 \cdot R} \right]$$

$$\boxed{\frac{V}{V_g} = \frac{D}{D'} \left\{ \frac{1 - \left(\frac{D'}{D}\right)\left(\frac{V_d}{V_g}\right)}{1 + \frac{(R_w + D \cdot R_{dson} + D' \cdot R_d)}{D'^2 R}} \right\} = \frac{D}{D'} \cdot \eta}$$

3.2 Buck-Boost Converter Efficiency & Gain Plots vs. R_{dson} in MOSFET

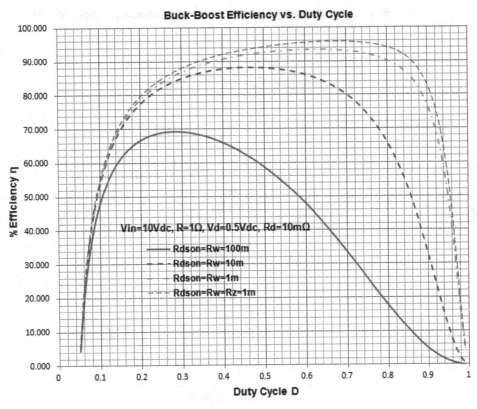

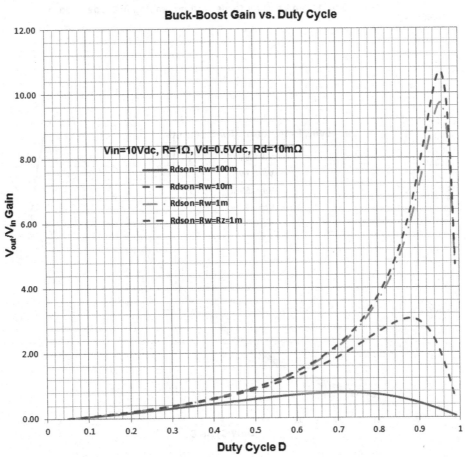

3.3 Buck-Boost Converter Efficiency & Gain Plots vs. Diode Voltage Drop

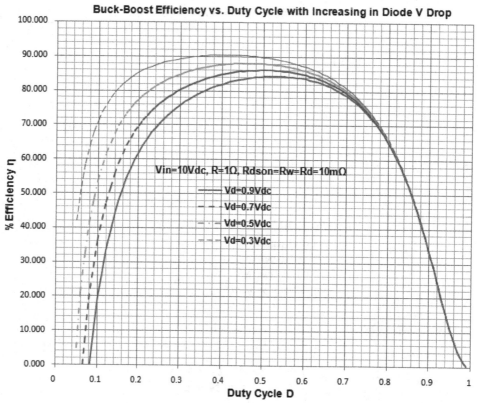

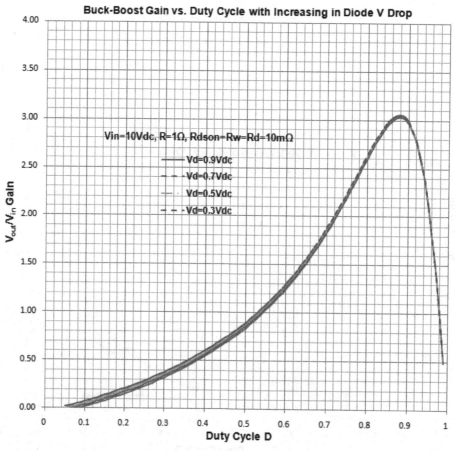

3.4 Buck-Boost Converter Efficiency & Gain Plots with Synchronous Rectifier

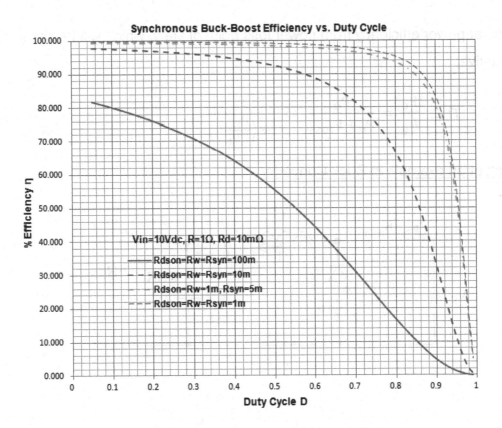

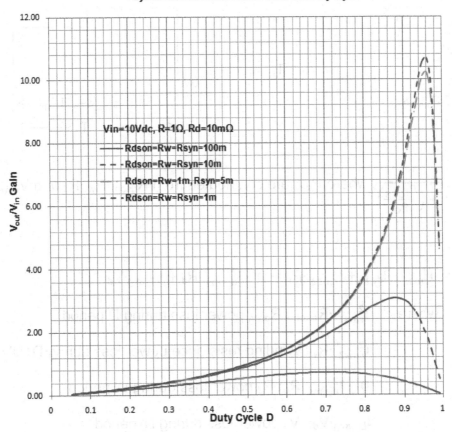

4. Power Efficiency Derivation from Power Loss Point of View

Let's look at power efficiency from power loss point of view, which will be simpler and more intuitive:

$$\eta \triangleq \frac{P_o}{P_g} = \frac{P_o}{P_o + P_{loss}} = \frac{1}{1 + \left(\frac{P_{loss}}{P_o}\right)}$$

<u>Let's look at an example in Buck Converter</u>

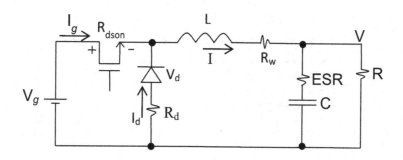

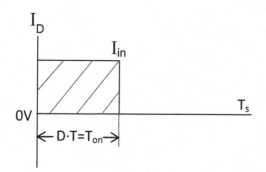
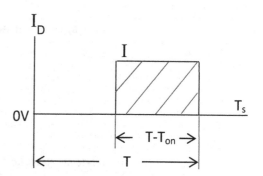

$$\eta = \frac{1}{1 + \left(\frac{P_{loss}}{P_o}\right)},$$ where P_{loss} are contributed from R_{dson}, R_w, R_d and V_d (ESR << R)

$$P_{loss} = I_{rms1}^2 \cdot R_{dson} + I_{rms2}^2 \cdot R_w + D' \cdot I_{d\_rms}^2 \cdot R_d + I_{D'\_avg} \cdot V_d$$

where $\mathbf{I_{rms1}^2 \cdot R_{dson}}$: R_{dson} power loss during D period

$\mathbf{I_{rms2}^2 \cdot R_w}$: Winding resistance power loss during D+D' periods

$\mathbf{D' \cdot I_{d\_rms}^2 \cdot R_d}$: R_d power loss during D' period

$\mathbf{I_{D'\_avg} \cdot V_d}$: V_d power loss during D' period

$$I_{rms1}^2 \triangleq \left(\frac{1}{T_s}\int_0^T i(t_1)^2 dt\right) = \frac{1}{T_s}\left(\int_0^{T_{on}} I^2 dt\right) + \left(\int_{T_{on}}^T (0)^2 dt\right) \quad I_{rms1}^2 \triangleq \left(\frac{1}{T_s}\int_0^T i(t_1)^2 dt\right)$$

$$I_{rms1}^2 \triangleq \frac{T_{on}}{T_s}\cdot I^2 = D\cdot I^2$$

Similarly,

$$I_{rms2}^2 = I^2$$

$$I_{D'\_rms}^2 = D'I^2$$

$I_{D'\_avg} = D' \cdot I$, where I is the average source current to resistive load R.

Therefore, $P_{loss} = D\cdot I^2 \cdot R_{dson} + I^2\cdot R_w + D' \cdot I^2\cdot R_d + D' \cdot I \cdot V_d$

$\quad P_o = I^2 \cdot R$

Therefore, $\quad \eta = \dfrac{1}{1+\left(\dfrac{P_{loss}}{Po}\right)} = \dfrac{1}{1+\left(\dfrac{I^2[D\cdot R_{dson} + D'\cdot R_d + R_w] + D'\cdot I\cdot V_d}{I^2\cdot R}\right)}$

$$\eta = \dfrac{1}{1+\left(\dfrac{[D\cdot R_{dson} + D'\cdot R_d + R_w]}{R} + \dfrac{D'\cdot V_d}{I\cdot R}\right)}$$

But $I\cdot R = V = D\cdot V_g\cdot \eta$

Eq1: $\quad \eta = \dfrac{1}{1+\left(\dfrac{[D\cdot R_{dson} + D'\cdot R_d + R_w]}{R} + \dfrac{D'\cdot V_d}{D\cdot \eta\cdot V_g}\right)}$

But we had;

Eq2: $\quad \eta = \dfrac{\left[1 - \left(\dfrac{D'}{D}\right)\left(\dfrac{V_d}{V_g}\right)\right]}{\left[1 + \dfrac{(D\cdot R_{dson} + D'\cdot R_d + R_w)}{R}\right]}$

Question: Is **Eq1** same as **Eq2**?

Let's solve for η in **Eq1**.

$$\eta\left[1 + \frac{(D \cdot R_{dson} + D' \cdot R_d + R_w)}{R} + \frac{D' \cdot V_d}{D \cdot \eta \cdot V_g}\right] = 1$$

$$\eta + \eta\left[\frac{(D \cdot R_{dson} + D' \cdot R_d + R_w)}{R}\right] + \frac{D' \cdot V_d}{D \cdot V_g} = 1$$

$$\eta\left[1 + \frac{(D \cdot R_{dson} + D' \cdot R_d + R_w)}{R}\right] = 1 - \frac{D' \cdot V_d}{D \cdot V_g}$$

$$\boxed{\eta = \frac{\left(1 - \left(\frac{D'}{D}\right)\left(\frac{V_d}{V_g}\right)\right)}{\left[1 + \frac{(D \cdot R_{dson} + D' \cdot R_d + R_w)}{R}\right]}}$$

Answer: **Eq1** is same as **Eq2**!!!!

Therefore, all we need is the RMS and Average currents for any converter and plug those into the power efficiency equation to get our transfer ratio between output and input voltage. Let us derive all the RMS and Average current waveforms that we would come across in any switch mode converter.

5. Common Waveforms for I^2_{rms} and I_{avg} for any Switch Mode Converters

I_{avg} **Definition:** $\boxed{I_{avg} \triangleq \dfrac{1}{T}\int_0^T i(t)\,dt}$

I^2_{rms} **Definition:** $\boxed{I^2_{rms} \triangleq \dfrac{1}{T}\int_0^T (i(t))^2\,dt}$

Common Waveforms	I^2_{rms}	I_{avg}
i(t) — DC Current	$I^2_{rms} = I^2$	$I_{avg} = I$
i(t)	$I^2_{rms} = D \cdot I^2$	$I_{avg} = D \cdot I$

Waveform	RMS	Average
	$I_{rms}^2 = D' \cdot I^2$	$I_{avg} = D' \cdot I$
	$I_{rms}^2 = I^2 \left[1 + \frac{1}{12} \left(\frac{\Delta I}{I} \right)^2 \right]$	$I_{avg} = I$
	$I_{rms}^2 = \frac{I_{pk}^2}{2}$	$I_{avg} = 0$

	$$I_{rms}^2 = D \cdot I^2 \left[1 + \frac{1}{12} \left(\frac{\Delta I}{I} \right)^2 \right]$$ (See Appendix A-1 for derivation)	$$I_{avg} = D \cdot I$$
	$$I_{rms}^2 = D' \cdot I^2 \left[1 + \frac{1}{12} \left(\frac{\Delta I}{I} \right)^2 \right]$$	$$I_{avg} = D' \cdot I$$
	$$I_{rms}^2 = \left[\frac{D+D_2}{3} \right] I^2_{PK}$$	$$I_{avg} = \left[\frac{D+D_2}{2} \right] I_{PK}$$

	$$I_{rms}^2 = \left[\frac{D}{3}\right] I_{PK}^2$$	$$I_{avg} = \left[\frac{D}{2}\right] I_{PK}$$
	$$I_{rms}^2 = \left[\frac{\Delta I^2}{12}\right]$$	If area_A = area_B, then: $$I_{avg} = 0$$
	$$I_{rms}^2 = \left[\frac{D}{3}\right] I_{PK}^2$$	$$I_{avg} = \left[\frac{D_2}{2}\right] I_{PK}$$

Example: find current through the Capacitor (I_{RC_rms}) for the following waveform pertaining to the circuit below.

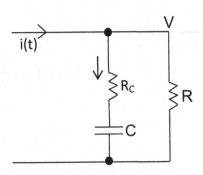

$$I_{RC}(t) = I(t) - V/R$$

$$I_{RC\_rms}^2 = D_2 \left(\frac{I_{pk}^2}{3} - I_{pk} \cdot I_{avg} + I_{avg}^2 \right)$$

Boost I_{rms}^2 current R_{ESR}

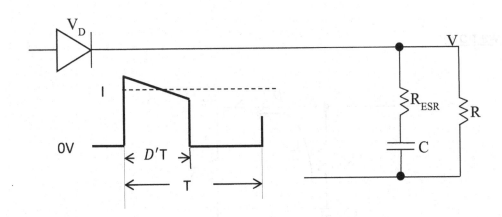

There can be no net DC current in the capacitor in steady state.

So DC or AVERAGE current of the diode must flow through Resistor, R.

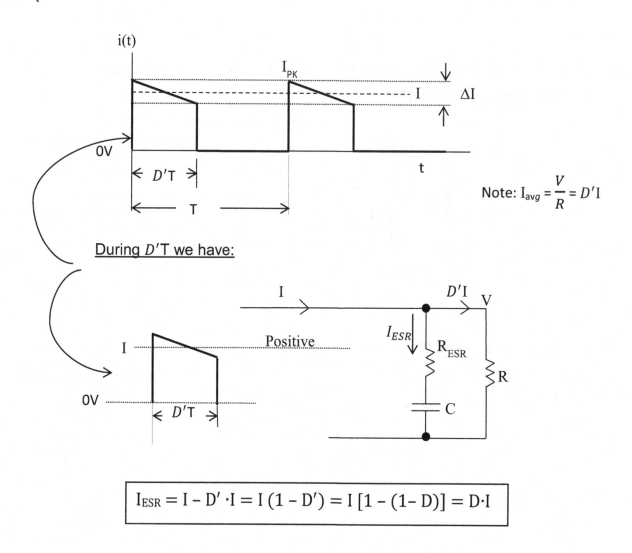

Note: $I_{avg} = \dfrac{V}{R} = D'I$

During D'T we have:

$$I_{ESR} = I - D' \cdot I = I\,(1 - D') = I\,[1 - (1 - D)] = D \cdot I$$

Now during D'T we have:

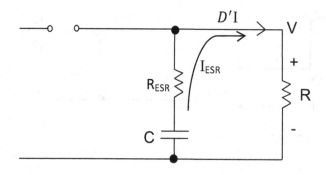

The capacitor must supply the AVERAGE current, V/R

But V/R = $D' \cdot I$ = I_{ESR}

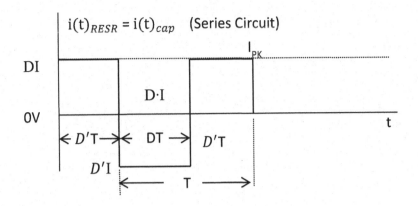

$$I^2_{rms} = D'(D \cdot I)^2 + D\,(D' \cdot I)^2$$

$$I^2_{rms} = D' \cdot D^2 \cdot I^2 + D \cdot D'^2 \cdot I^2 = D \cdot D' \cdot I^2\,(D + D') = D \cdot D' \cdot I^2$$

Therefore, the RMS current into capacitor and R_{ESR} resistor is:

$$\boxed{I^2_{rms} = D \cdot D' \cdot I^2}$$

Hence, the power dissipated in the R_{ESR} is:

$$\boxed{P_{ESR} = I^2_{rms} \cdot R_{ESR} = D \cdot D' \cdot I^2 \cdot R_{ESR}}$$

6. Find Q of LCR Network

6.1 The Hard Way to Find Q of the LCR Network

Finding the Q in LCR circuits is important because a high Q can lead to instability in closed loop control of converter or instability when adding an input filter to a converter.

Finding Z_{OL}:

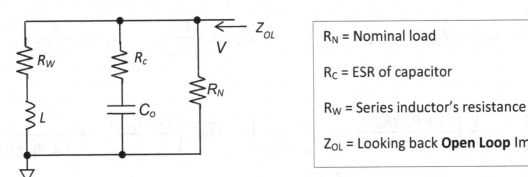

R_N = Nominal load

R_C = ESR of capacitor

R_W = Series inductor's resistance

Z_{OL} = Looking back **Open Loop** Impedance

$$Z_{OL} = \frac{1}{\frac{1}{R_N} + \left(\frac{s \cdot C_O}{s \cdot C_O \cdot R_C + 1} + \frac{1}{s \cdot L + R_W}\right)} = \frac{R_N(sC_O R_C + 1)(s \cdot L + R_W)}{(sC_O R_C + 1)(s \cdot L + R_W) + R_N \cdot s \cdot C_O(s \cdot L + R_W) + R_N(s \cdot C_O \cdot R_C + 1)}$$

$$= \frac{R_N(s \cdot C_O R_C + 1)(s \cdot L + R_W)}{s^2 \cdot L \cdot C_O[R_C + R_N] + s\left[C_O \cdot R_C \cdot R_W + L + C_O \cdot R_N \cdot R_W + C_O R_N \cdot R_C\right] + [R_W + R_N]}$$

$$= \frac{R_N \cdot R_W}{(R_W + R_N)} \cdot \frac{(s \cdot C_O R_C + 1)\left(s \cdot \frac{L}{R_W} + 1\right)}{s^2 \cdot L \cdot C_O\left[\frac{R_C + R_N}{R_W + R_N}\right] + \frac{s \cdot L \cdot C_O}{(R_W + R_N)} \cdot \left[\frac{(R_C \cdot R_W + R_N \cdot R_W + R_N \cdot R_C)}{L} + \frac{1}{C_O}\right] + 1}$$

$$= (R_N \| R_W) \cdot \frac{(s \cdot C_O \cdot R_C + 1)\left(s \cdot \frac{L}{R_W} + 1\right)}{s^2 \cdot L \cdot C_O\left[\frac{R_C + R_N}{R_W + R_N}\right] + s \cdot L \cdot C_O\left[\frac{(R_C + R_N \| R_W)}{L} + \frac{1}{C_O(R_W + R_N)}\right] + 1}$$

Let $\omega_O \triangleq \frac{1}{\sqrt{LC_O}}$, also note:

$$R_N \gg R_C$$

$$R_N \gg R_W$$

$$\Longrightarrow \quad \therefore \frac{R_C + R_N}{R_W + R_N} \approx \frac{R_N}{R_N} = 1$$

$$Z_{OL} \cong (R_N \| R_W) \cdot \frac{(s \cdot C_O \cdot R_C + 1)\left(s \cdot \frac{L}{R_W} + 1\right)}{\left[\frac{s}{\omega_O}\right]^2 + \frac{s}{\omega_O{}^2}\left[\frac{(R_C + R_N \| R_W)}{L} + \frac{1}{C_O(R_W + R_N)}\right] + 1}$$

The **EASY** way: General form…

$$Z_{OL} = R_{dc} \cdot \frac{\left(\frac{s}{\omega_{Z1}} + 1\right)\left(\frac{s}{\omega_{Z2}} + 1\right)}{\left[\frac{s}{\omega_0}\right]^2 + \frac{1}{Q} \cdot \frac{s}{\omega_0} + 1}$$

$$R_{dc} = R_N \| R_W, \qquad \omega_{Z1} = \frac{1}{C_0 \cdot R_c}, \qquad \omega_{Z2} = \frac{1}{L/R_W}, \qquad \omega_0 = \frac{1}{\sqrt{LC_0}}$$

$$\frac{1}{Q} = \frac{1}{\omega_0}\left[\frac{R_C + R_W\|R_N}{L} + \frac{1}{C_0(R_W + R_N)}\right] = \sqrt{LC_0}\left[\frac{R_C + R_W\|R_N}{L} + \frac{1}{C_0(R_W + R_N)}\right]$$

$$= \frac{R_C + R_W\|R_N}{\sqrt{L/C_0}} + \frac{\sqrt{L/C_0}}{(R_W + R_N)}$$

Let $R_o \triangleq \sqrt{\dfrac{L}{C_0}}$

$$\frac{1}{Q} = \frac{R_C + R_W\|R_N}{R_0} + \frac{R_0}{(R_W + R_N)} = \frac{1}{Q_c} + \frac{1}{Q_L} \text{ , where } Q = Q_c\|Q_L$$

$$\frac{1}{Q_c} \triangleq \frac{R_C + R_W\|R_N}{R_0} \quad \rightarrow \quad Q_C = \frac{R_0}{R_C + R_W\|R_N}$$

$$\frac{1}{Q_L} \triangleq \frac{R_0}{(R_W + R_N)} \quad \rightarrow \quad Q_L = \frac{R_W + R_N}{R_0}$$

6.2 The short way to Find Q of the RLC Network:

Finding Z_{OL}:

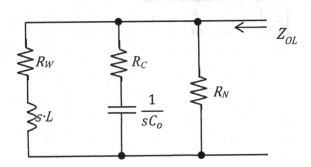

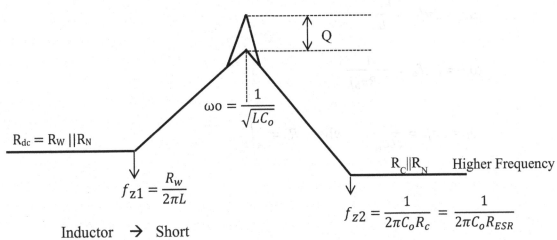

$$\omega o = \frac{1}{\sqrt{LC_o}}$$

$R_{dc} = R_W \,||\, R_N$

$$f_{z1} = \frac{R_w}{2\pi L}$$

$R_c || R_N$ Higher Frequency

$$f_{z2} = \frac{1}{2\pi C_o R_c} = \frac{1}{2\pi C_o R_{ESR}}$$

Inductor → Short

Capacitor → Open

Low frequency or DC

General Form:

$$Z_{OL} = (R_N \| R_w) \cdot \frac{\left(\frac{s}{\omega_{z1}} + 1\right)\left(\frac{s}{\omega_{z2}} + 1\right)}{\left[\frac{s}{\omega_o}\right]^2 + \frac{1}{Q} \cdot \frac{s}{\omega_o} + 1}$$

$$Q = Q_C \| Q_L$$

$$Q_L = \frac{R_W + R_N}{R_o}$$

$$Q_C = \frac{R_o}{R_c + R_w \| R_N}$$

$$\omega_{Z1} = 2\pi f_{z1} = \frac{R_W}{L}$$

$$\omega_{Z2} = 2\pi f_{z2} = \frac{1}{C_o R_{ESR}}$$

$$\omega_O = 2\pi f_o = \frac{1}{\sqrt{LC_o}}, \quad \text{where } R_o = \sqrt{\frac{L}{C_o}}$$

6.3 Derive Transfer Function and Que (Q) of a Low Pass Filter - the Easy Way

Find Q_e (overall Q) of any single LCR filters with or without a damping network:

General Procedure

Step 1: Find K (DC term or Gain) by shorting all L's and opening all C's in a filter network.

Then find:

$$K = \frac{V_o}{V_{in}}$$

Step 2: Find equivalent circuit near frequency of interest; $f = f_o$.

Where: $f_o = \dfrac{1}{2\pi\sqrt{LC}}$

Step 3: Find $Q_{eqC} = \dfrac{R_o}{R_{eqC}}$, which is capacitor's Que.

$$Q_{eqC} \triangleq \frac{R_o}{R_{eqC}} \text{ , where Characteristic impedance for } R_o = \sqrt{\frac{L}{C}}$$

To find Q_c: "short" the power input and all inductors, which C "sees".

Solve for Equivalent Resistance R_{eqC} in parallel and/or in series with C "sees".

Step 4: Find Q_L which is the inductor's Que.

$$Q_{eqL} \triangleq \frac{R_{eqL}}{R_o}$$

To find Q_L: "short" the power input and "open" all Capacitors, which L "sees".

Find Equivalent Resistance R_{eqL} in parallel and/or in series with L "sees".

Step 5: Note that the Que of a system, Q_e, is determined by $Q_c \| Q_L$.

$$\frac{1}{Q_e} = \frac{1}{Q_c} + \frac{1}{Q_L}$$

Step 6: Write equation in the general form:

$$H(s) = \frac{K}{\left(\frac{s}{\omega_o}\right)^2 + \frac{1}{Q_e} \cdot \left(\frac{s}{\omega_o}\right) + 1} \quad , \text{ where } \omega_{o.} = \frac{1}{\sqrt{LC}}$$

$$\text{Note: } Q_e = \frac{Q_c Q_L}{Q_c + Q_L} \quad \text{ and } \quad \omega_o = 2\pi f_o$$

Example #1: calculate Que in the following circuit.

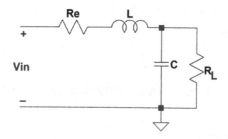

Step 1: Find K: short L's and Open C's.

$$\rightarrow K = \frac{V_o}{V_{in}} = \frac{R_L}{R_L + R_e}$$

Step 2: Find equivalent circuit near frequency of interest $f = f_o$.

$$\rightarrow f_{o.} = \frac{1}{2\pi\sqrt{LC}}$$

Step 3: Find $Q_{eqC} = \dfrac{R_o}{R_{eqC}}$ shorting the input and the inductor L, which C "sees".

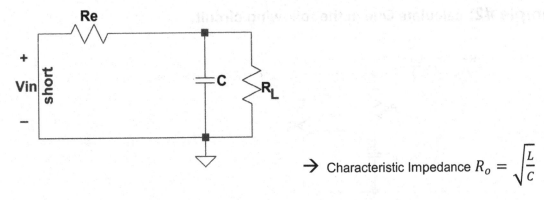

→ Characteristic Impedance $R_o = \sqrt{\dfrac{L}{C}}$

where $R_{eqC} = R_e \| R_L$ and $Q_{eqC} \triangleq \dfrac{R_o}{R_{eqC}} = \dfrac{R_o}{R_e \| R_L}$

Step 4: Find Q_L by shorting the input and opening all capacitors, which L "sees".

→ $Q_{eqL} \triangleq \dfrac{R_{eqL}}{R_o} = \dfrac{R_e + R_L}{R_o}$

Step 5: Calculate the system Que, Q_e:

$$\frac{1}{Q_e} = \frac{1}{Q_c} + \frac{1}{Q_L}$$

$$H(s) = \frac{K}{\left(\dfrac{s}{\omega_o}\right)^2 + \dfrac{1}{Q_e}\left(\dfrac{1}{\omega_o}\right) + 1}$$

where $\omega_o = \dfrac{1}{\sqrt{LC}}$ and $K = \dfrac{R_L}{R_L + R_e}$

Example #2: calculate Que in the following circuit.

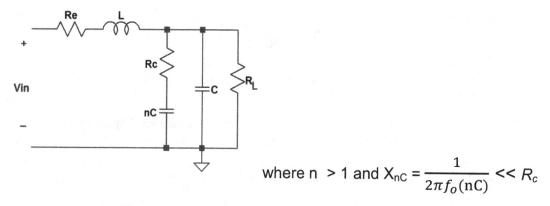

where n > 1 and $X_{nC} = \dfrac{1}{2\pi f_o(nC)} \ll R_c$

Step 1: Find K: short L's and Open C's.

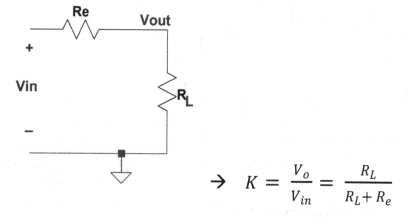

$\rightarrow \quad K = \dfrac{V_o}{V_{in}} = \dfrac{R_L}{R_L + R_e}$

Step 2: Draw equivalent circuit near frequency of interest $f = f_o$.

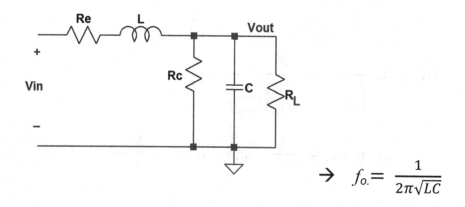

$\rightarrow \quad f_o = \dfrac{1}{2\pi\sqrt{LC}}$

Step 3: Find $Q_{eqC} = Q_C = \dfrac{R_o}{R_{eqC}}$ by shorting the power input and all inductor L's.

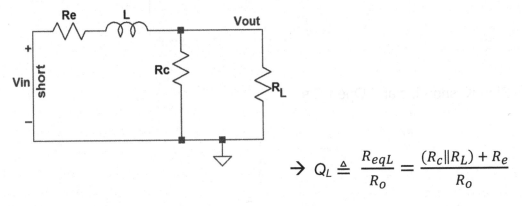

→ Characteristic Impedance $R_o = \sqrt{\dfrac{L}{C}}$

where $R_{eqC} = R_e \| R_c \| R_L$ and $Q_C \triangleq \dfrac{R_o}{R_{eqC}} = \dfrac{R_o}{R_e \| R_c \| R_L}$

Step 4: Determine inductor's Q_L by opening all capacitors and shorting input, which L "sees".

→ $Q_L \triangleq \dfrac{R_{eqL}}{R_o} = \dfrac{(R_c \| R_L) + R_e}{R_o}$

Step 5: Calculate the system Que, Q_e:

$$\frac{1}{Q_e} = \frac{1}{Q_c} + \frac{1}{Q_L} = \frac{R_e \| R_c \| R_L}{R_o} + \frac{R_o}{(R_c \| R_L) + R_e}$$

where $Q_e = \dfrac{Q_c \cdot Q_L}{Q_c + Q_L}$ and note that $Q_L \ll Q_c$

Hence $Q_e \cong Q_L = \dfrac{(R_c \| R_L) + R_e}{R_o}$

Step 6: Write the equation:

$$H(s) = \frac{K}{\left(\frac{s}{\omega_o}\right)^2 + \frac{1}{Q_e}\left(\frac{1}{\omega_o}\right) + 1}$$

$$\text{where } \omega_o = \frac{1}{\sqrt{LC}} \quad \text{and} \quad K = \frac{R_L}{R_L + R_e}$$

Example #3: calculate Que in the following circuit.

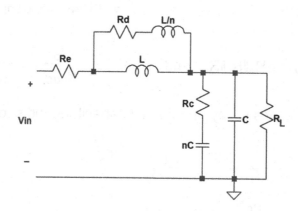

Step 1: Find K: short L's and Open C's.

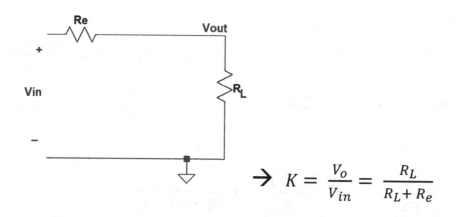

$$\rightarrow K = \frac{V_o}{V_{in}} = \frac{R_L}{R_L + R_e}$$

Step 2: Draw equivalent circuit near frequency of interest $f = f_o.$

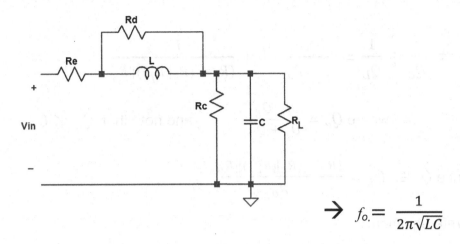

$$\rightarrow \quad f_{o.} = \frac{1}{2\pi\sqrt{LC}}$$

Step 3: Find $Q_{eqC} = Q_C = \dfrac{R_o}{R_{eqC}}$ by shorting the input and all inductor L's.

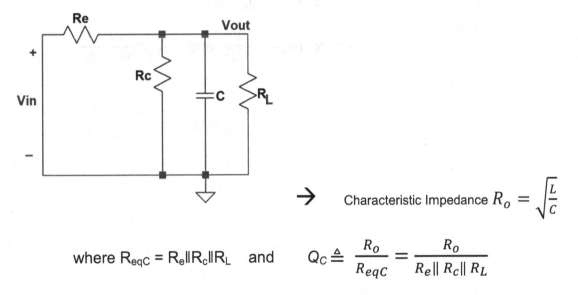

$$\rightarrow \quad \text{Characteristic Impedance } R_o = \sqrt{\frac{L}{C}}$$

where $R_{eqC} = R_e \| R_c \| R_L$ and $Q_C \triangleq \dfrac{R_o}{R_{eqC}} = \dfrac{R_o}{R_e \| R_c \| R_L}$

Step 4: Find inductor's Q_L by shorting input V_{in} and opening all capacitors, which L "sees".

$$\rightarrow \quad Q_L \triangleq \frac{R_{eqL}}{R_o} = \frac{(R_e + (R_c \| R_L)) \| R_d}{R_o}$$

Step 5: Calculate the system Que, Q_e:

$$\frac{1}{Q_e} = \frac{1}{Q_C} + \frac{1}{Q_L} = \frac{R_e \| R_c \| R_L}{R_o} + \frac{R_o}{(R_e + (R_c \| R_L)) \| R_d}$$

where $Q_e = \dfrac{Q_c \cdot Q_L}{Q_c + Q_L}$, and note that $Q_L \ll Q_C$

Hence $Q_e \cong Q_L = \dfrac{(R_e + (R_c \| R_L)) \| R_d}{R_o}$

Step 6: Write equation:

$$H(s) = \frac{K}{\left(\dfrac{s}{\omega_o}\right)^2 + \dfrac{1}{Q_e}\left(\dfrac{1}{\omega_o}\right) + 1}$$

where $\omega_o = \dfrac{1}{\sqrt{LC}}$ and $K = \dfrac{R_L}{R_L + R_e}$

6.4 Few Examples of Finding Q_e Using the above Technique

The purpose of this section to find an easier way to evaluate a passive Low Pass Filter (LPF), which would enable us to obtain a transfer function with few equations.

Common Questions:

1. If a LPF is given, how do we determine the peaking of the LPF (i.e. the Q of the filter)?
2. If the Q is too high, how do we lower the Q without affecting the rest of the filter characteristics?
3. If the filter is complex, how can we simplify the complexity without losing any functionality?

Answers:

We will start with a general form of second order LPF and add additional poles and zeros to the filter. An objective is to obtain a transfer function of a complex filter by sight. The actual Q value of a passive filter may be verified by LTSpice simulation software, free download from Linear Technology/Analog Devices webpage:

https://www.analog.com/en/design-center/design-tools-and-calculators/ltspice-simulator.html).

In addition, more learning events via seminars and/ Youtube videos can be found in the following educational website and video links:

http://smartpowertechnology.net/

https://www.youtube.com/watch?v=AlHtQBy_2pY

https://www.youtube.com/watch?v=R-xjAEY0Lug

https://www.youtube.com/watch?v=N_YkjR7wBzk

https://www.youtube.com/watch?v=lxiD6Qc003g

https://www.youtube.com/watch?v=VAu52H9gcjs

https://www.youtube.com/watch?v=fNktac3jKxs

https://www.youtube.com/watch?v=3QCGPxKmnQI

https://www.youtube.com/watch?v=h2r8-vXp-s0

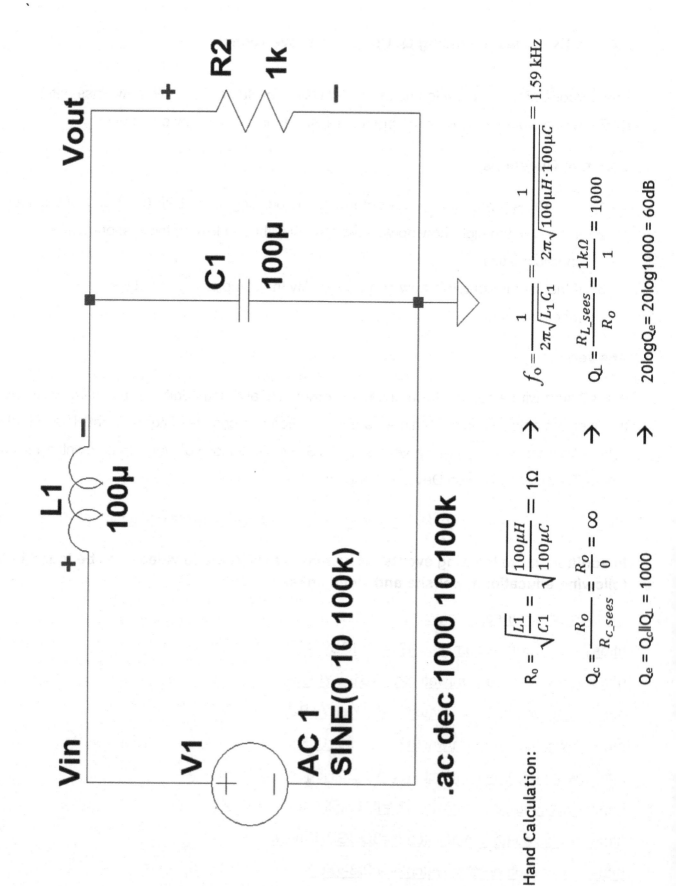

Hand Calculation:

$$R_o = \sqrt{\frac{L1}{C1}} = \sqrt{\frac{100\mu H}{100\mu C}} = 1\Omega$$

$$f_o = \frac{1}{2\pi\sqrt{L_1 C_1}} = \frac{1}{2\pi\sqrt{100\mu H \cdot 100\mu C}} = 1.59 \text{ kHz}$$

$$Q_c = \frac{R_o}{R_{C\_sees}} = \frac{R_o}{0} = \infty$$

$$Q_L = \frac{R_{L\_sees}}{R_o} = \frac{1k\Omega}{1} = 1000$$

$$Q_e = Q_c \| Q_L = 1000$$

$$20 \log Q_e = 20 \log 1000 = 60 dB$$

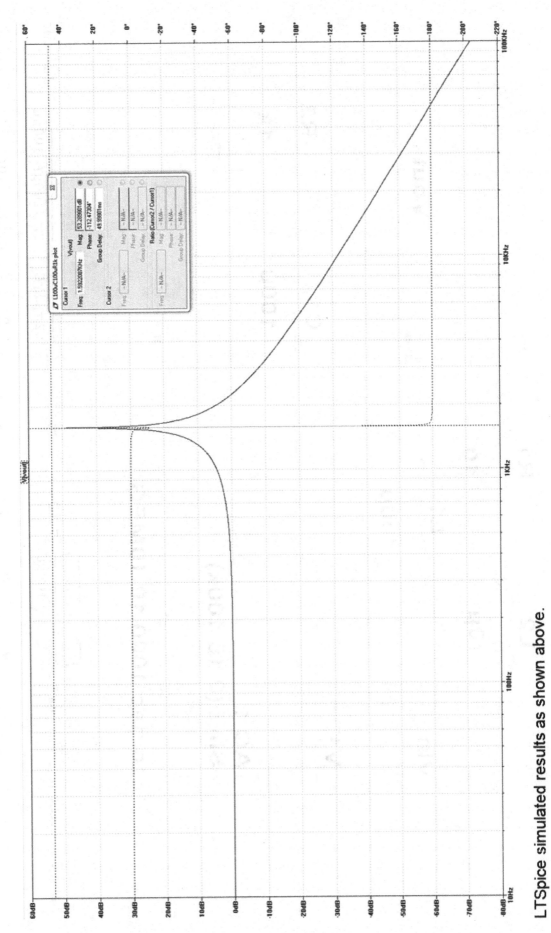

LTSpice simulated results as shown above.

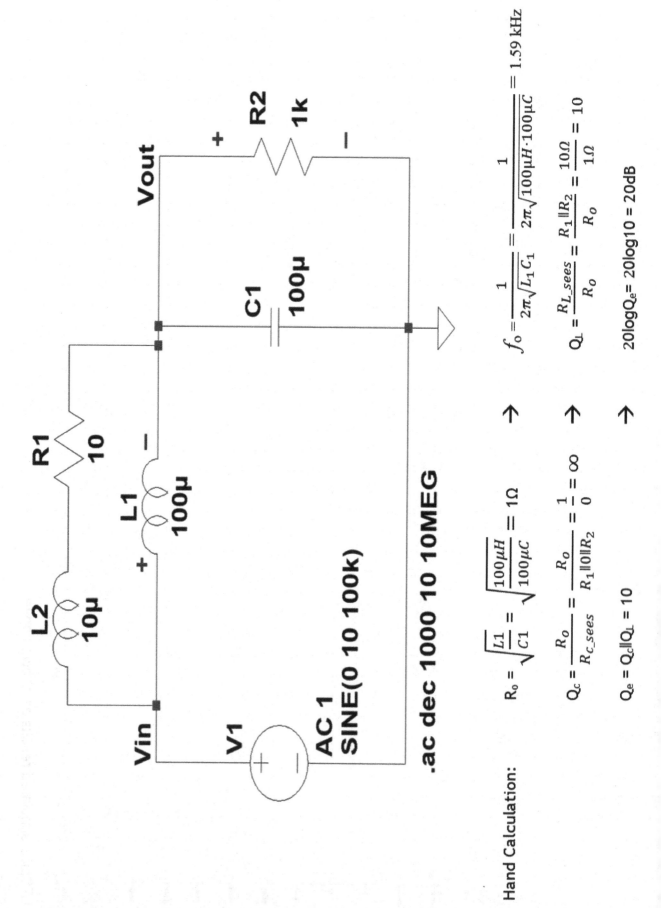

Hand Calculation:

$$R_o = \sqrt{\frac{L1}{C1}} = \sqrt{\frac{100\mu H}{100\mu C}} = 1\Omega$$

$$f_o = \frac{1}{2\pi\sqrt{L_1 C_1}} = \frac{1}{2\pi\sqrt{100\mu H \cdot 100\mu C}} = 1.59 \text{ kHz}$$

$$Q_c = \frac{R_o}{R_{c\_sees}} = \frac{R_o}{R_1\|0\|R_2} = \frac{1}{0} = \infty$$

$$Q_L = \frac{R_{L\_sees}}{R_o} = \frac{R_1\|R_2}{R_o} = \frac{10\Omega}{1\Omega} = 10$$

$$Q_e = Q_c\|Q_L = 10$$

$$20\log Q_e = 20\log 10 = 20\text{dB}$$

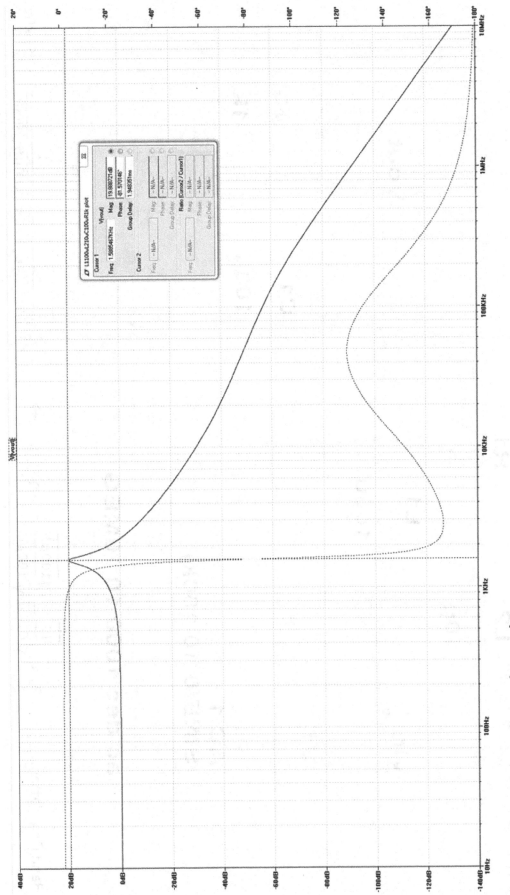

LTSpice simulated results as shown above.

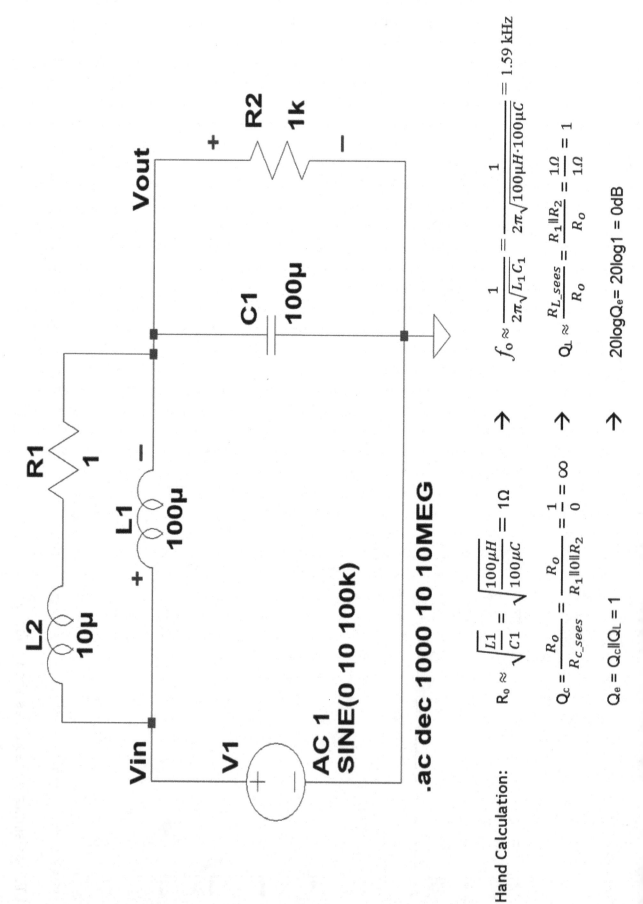

Hand Calculation:

$$R_o \approx \sqrt{\frac{L1}{C1}} = \sqrt{\frac{100\mu H}{100\mu C}} = 1\Omega$$

$$f_o \approx \frac{1}{2\pi\sqrt{L_1 C_1}} = \frac{1}{2\pi\sqrt{100\mu H \cdot 100\mu C}} = 1.59\ kHz$$

$$Q_c = \frac{R_o}{R_{c\_sees}} = \frac{R_o}{R_1\|0\|R_2} = \frac{1}{0} = \infty$$

$$Q_L \approx \frac{R_{L\_sees}}{R_o} = \frac{R_1\|R_2}{R_o} = \frac{1\Omega}{1\Omega} = 1$$

$$Q_e = Q_c\|Q_L = 1$$

$$20\log Q_e = 20\log 1 = 0dB$$

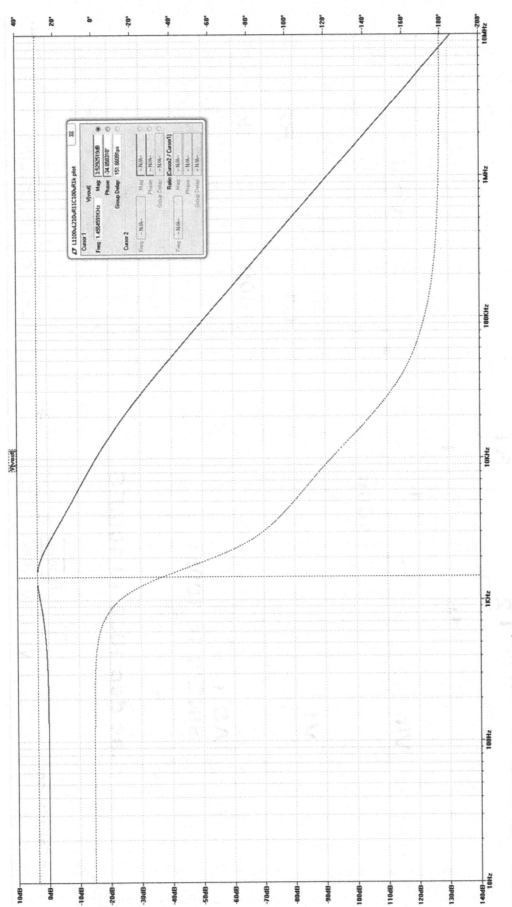

LTSpice simulated results as shown above.

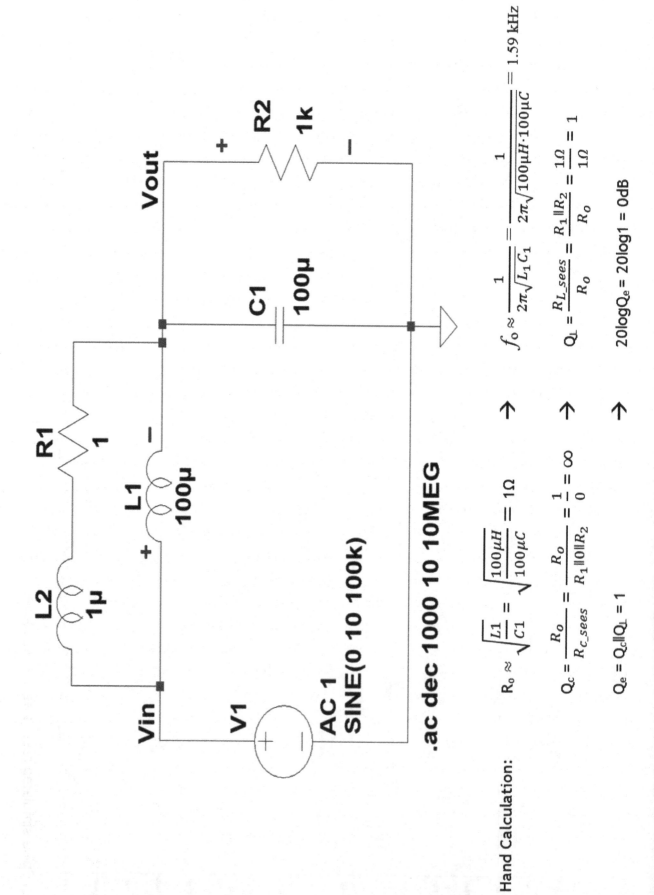

Hand Calculation:

$$R_o \approx \sqrt{\frac{L1}{C1}} = \sqrt{\frac{100\mu H}{100\mu C}} = 1\Omega$$

$$f_o \approx \frac{1}{2\pi\sqrt{L_1 C_1}} = \frac{1}{2\pi\sqrt{100\mu H \cdot 100\mu C}} = 1.59\ kHz$$

$$Q_c = \frac{R_o}{R_{c\_sees}} = \frac{R_o}{R_1 \| 0 \| R_2} = \frac{1}{0} = \infty$$

$$Q_L = \frac{R_{L\_sees}}{R_o} = \frac{R_1 \| R_2}{R_o} = \frac{1\Omega}{1\Omega} = 1$$

$$Q_e = Q_c \| Q_L = 1$$

$$20 \log Q_e = 20 \log 1 = 0 dB$$

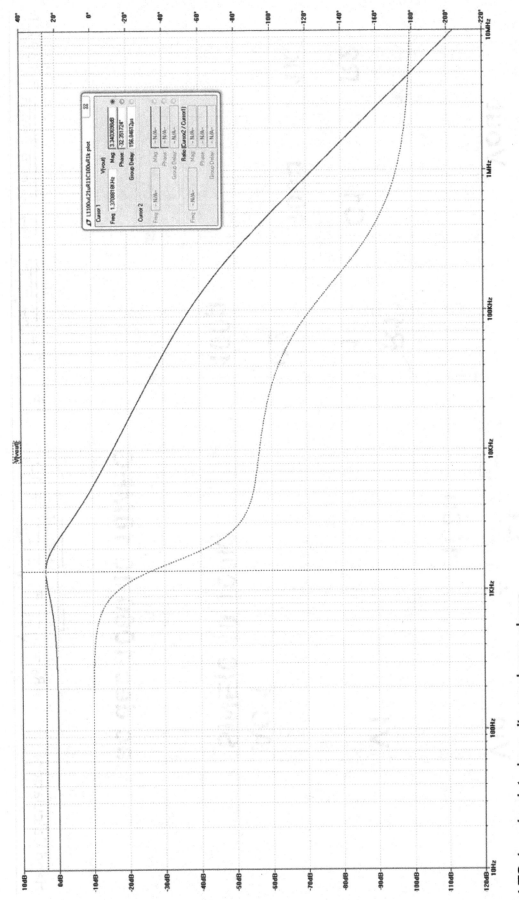

LTSpice simulated results as shown above.

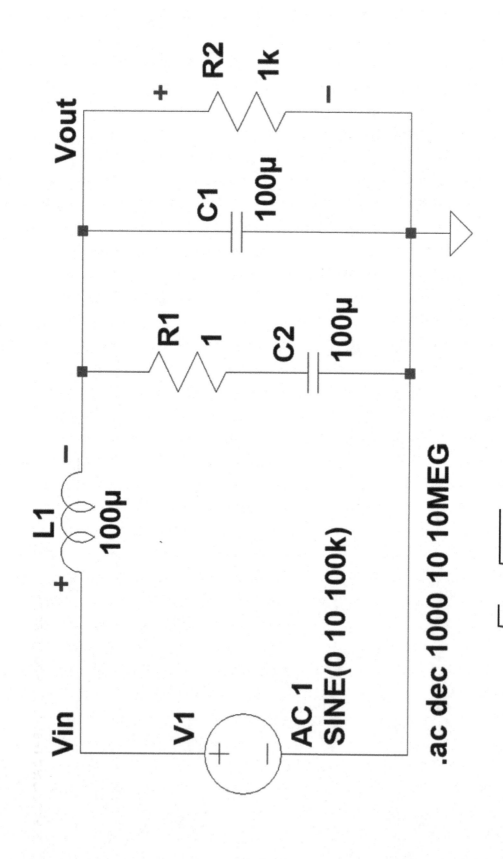

Hand Calculation:

$$R_o = \sqrt{\frac{L1}{C1}} = \sqrt{\frac{100\mu H}{100\mu C}} = 1\Omega \quad \rightarrow \quad f_o = \frac{1}{2\pi\sqrt{L_1 C_1}} = \frac{1}{2\pi\sqrt{100\mu H \cdot 100\mu C}} = 1.59\ kHz$$

$$Q_c = \frac{R_o}{R_{c\_sees}} = \frac{R_o}{0\|R_1\|R_2} = \frac{1}{0} = \infty \quad \rightarrow \quad Q_L = \frac{R_{L\_sees}}{R_o} = \frac{R_1\|R_2}{R_o} = \frac{1\Omega}{1\Omega} = 1$$

$$Q_e = Q_c\|Q_L = 1 \quad \rightarrow \quad 20\log Q_e = 20\log 1 = 0dB$$

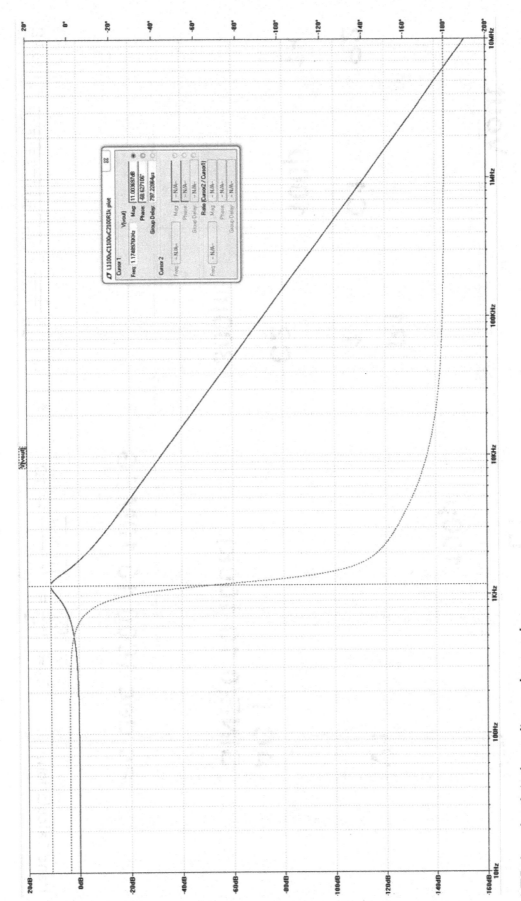

LTSpice simulated results as shown above.

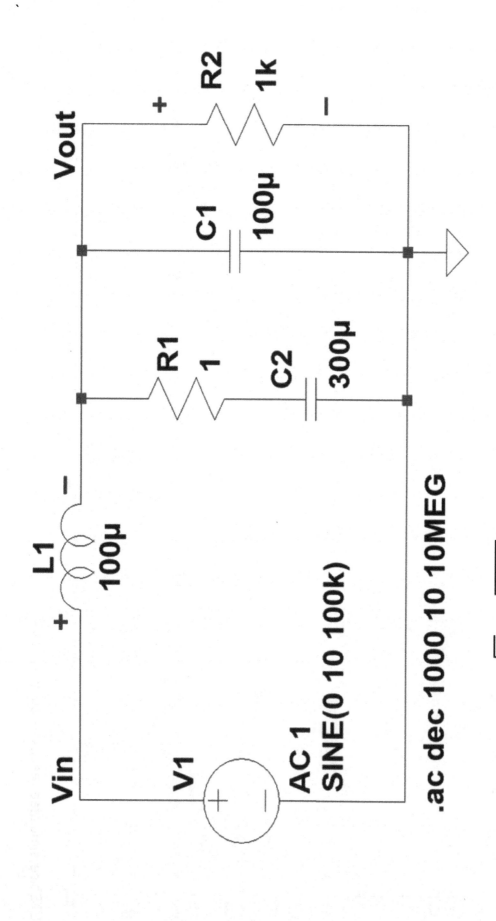

Hand Calculation:

$$R_o = \sqrt{\frac{L1}{C1}} = \sqrt{\frac{100\mu H}{100\mu C}} = 1\Omega \quad \rightarrow \quad f_o = \frac{1}{2\pi\sqrt{L_1 C_1}} = \frac{1}{2\pi\sqrt{100\mu H \cdot 100\mu C}} = 1.59 \text{ kHz}$$

$$Q_c = \frac{R_o}{R_{C\_sees}} = \frac{R_o}{R_1 \| R_2} = \frac{1}{0} = \infty \quad \rightarrow \quad Q_L = \frac{R_{L\_sees}}{R_o} = \frac{R_1 \| R_2}{R_o} = \frac{1\Omega}{1\Omega} = 1$$

$$Q_e = Q_c \| Q_L = 1 \quad \rightarrow \quad 20\log Q_e = 20\log 1 = 0\text{dB}$$

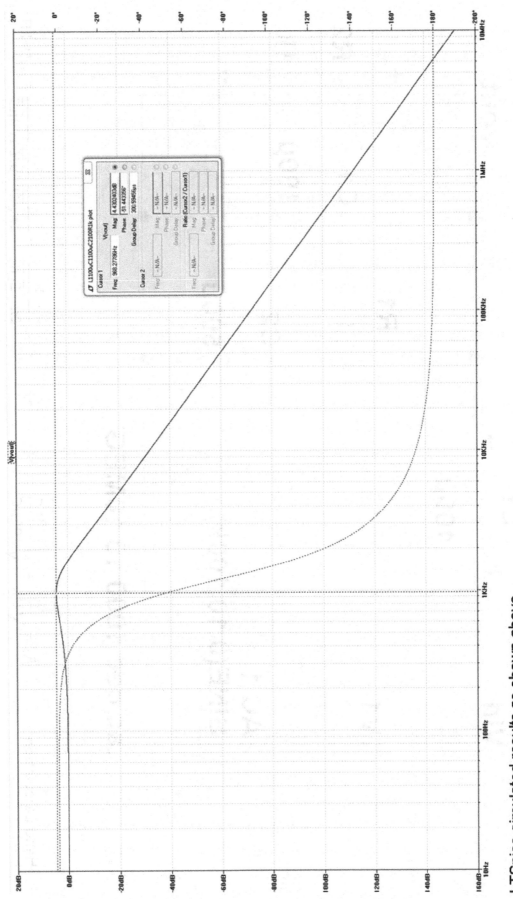

LTSpice simulated results as shown above.

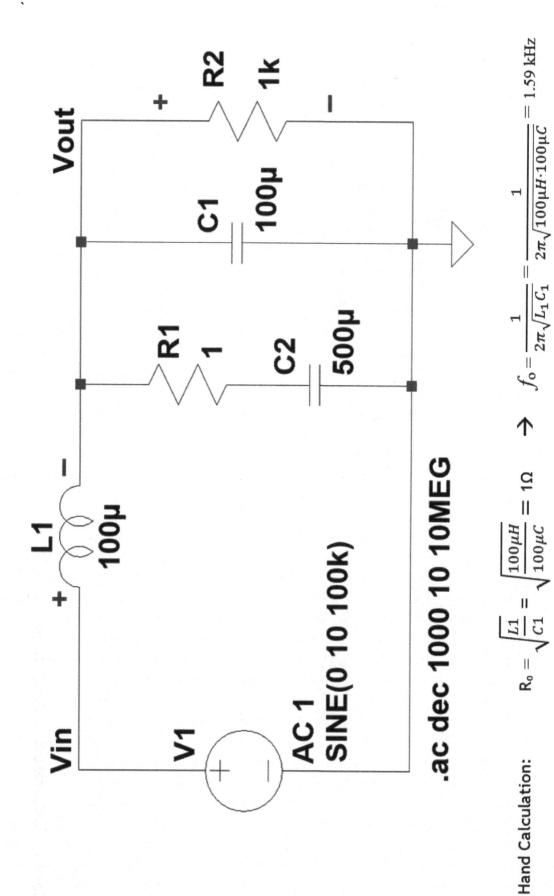

Hand Calculation:

$$R_o = \sqrt{\frac{L1}{C1}} = \sqrt{\frac{100\mu H}{100\mu C}} = 1\Omega \quad \rightarrow \quad f_o = \frac{1}{2\pi\sqrt{L_1 C_1}} = \frac{1}{2\pi\sqrt{100\mu H \cdot 100\mu C}} = 1.59 \text{ kHz}$$

$$Q_c = \frac{R_o}{R_{c\_sees}} = \frac{R_o}{0\|R_1\|R_2} = \frac{1}{0} = \infty \quad \rightarrow \quad Q_L = \frac{R_{L\_sees}}{R_o} = \frac{R_1\|R_2}{R_o} = \frac{1\Omega}{1\Omega} = 1$$

$$Q_e = Q_c\|Q_L = 1 \quad \rightarrow \quad 20\log Q_e = 20\log 1 = 0\text{dB}$$

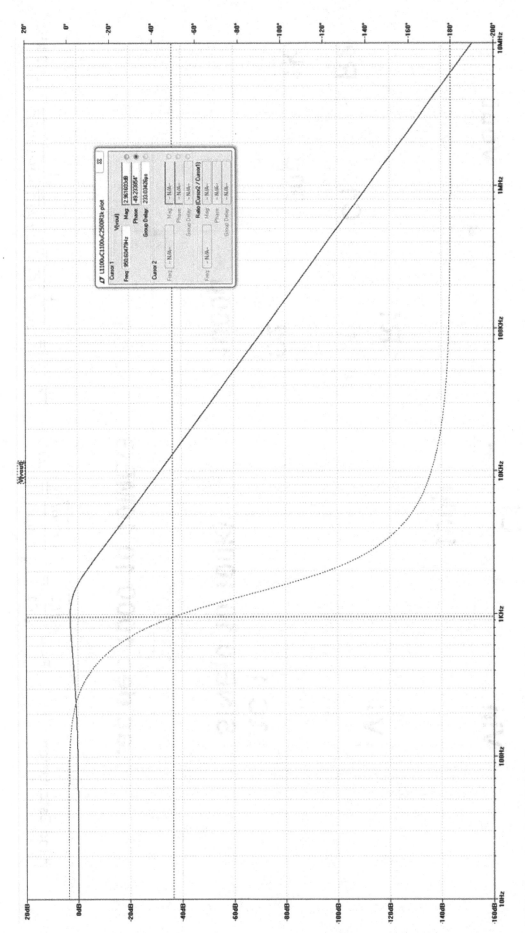

LTSpice simulated results as shown above.

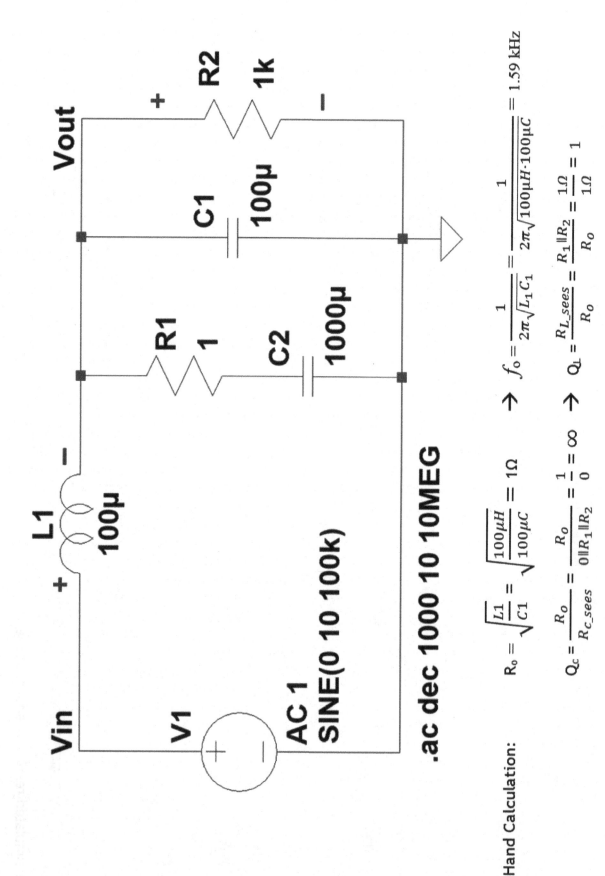

Hand Calculation:

$$R_o = \sqrt{\frac{L1}{C1}} = \sqrt{\frac{100\mu H}{100\mu C}} = 1\Omega \qquad \rightarrow \qquad f_o = \frac{1}{2\pi\sqrt{L_1 C_1}} = \frac{1}{2\pi\sqrt{100\mu H \cdot 100\mu C}} = 1.59\ kHz$$

$$Q_c = \frac{R_o}{R_{c\_sees}} = \frac{R_o}{0\|R_1\|R_2} = \frac{1}{0} = \infty \qquad \rightarrow \qquad Q_L = \frac{R_{L\_sees}}{R_o} = \frac{R_1\|R_2}{R_o} = \frac{1\Omega}{1\Omega} = 1$$

$$Q_e = Q_c\|Q_L = 1 \qquad \rightarrow \qquad 20logQ_e = 20log1 = 0dB$$

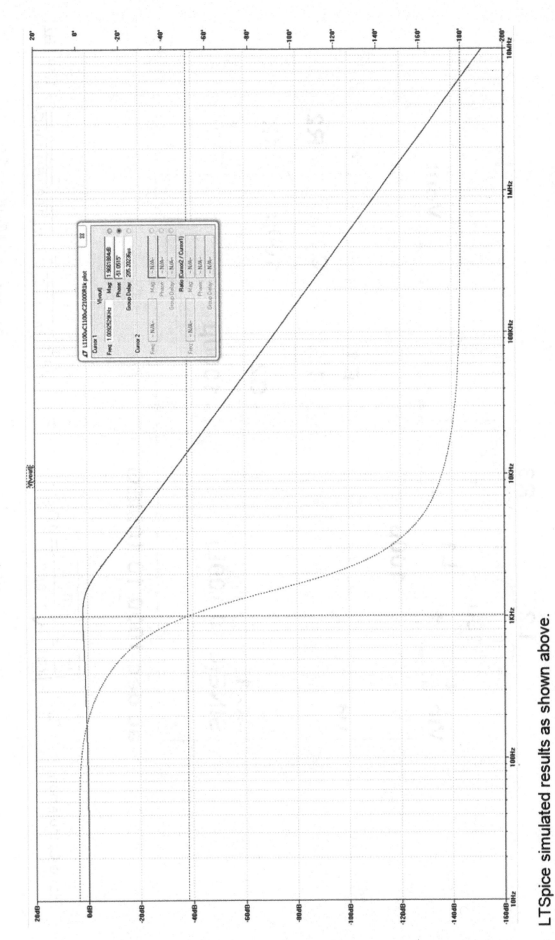

LTSpice simulated results as shown above.

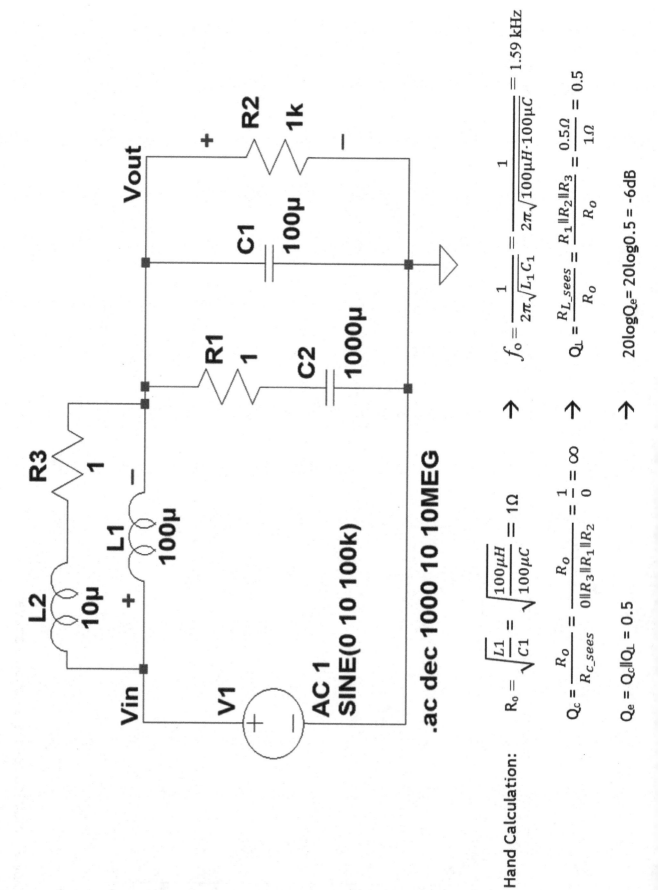

Hand Calculation:

$$R_o = \sqrt{\frac{L1}{C1}} = \sqrt{\frac{100\mu H}{100\mu C}} = 1\Omega$$

$$\rightarrow \quad f_o = \frac{1}{2\pi\sqrt{L_1 C_1}} = \frac{1}{2\pi\sqrt{100\mu H \cdot 100\mu C}} = 1.59 \text{ kHz}$$

$$Q_c = \frac{R_o}{R_{c\_sees}} = \frac{R_o}{0\|R_3\|R_1\|R_2} = \frac{1}{0} = \infty$$

$$\rightarrow \quad Q_L = \frac{R_{L\_sees}}{R_o} = \frac{R_1\|R_2\|R_3}{R_o} = \frac{0.5\Omega}{1\Omega} = 0.5$$

$$Q_e = Q_c\|Q_L = 0.5$$

$$\rightarrow \quad 20\log Q_e = 20\log 0.5 = -6 \text{dB}$$

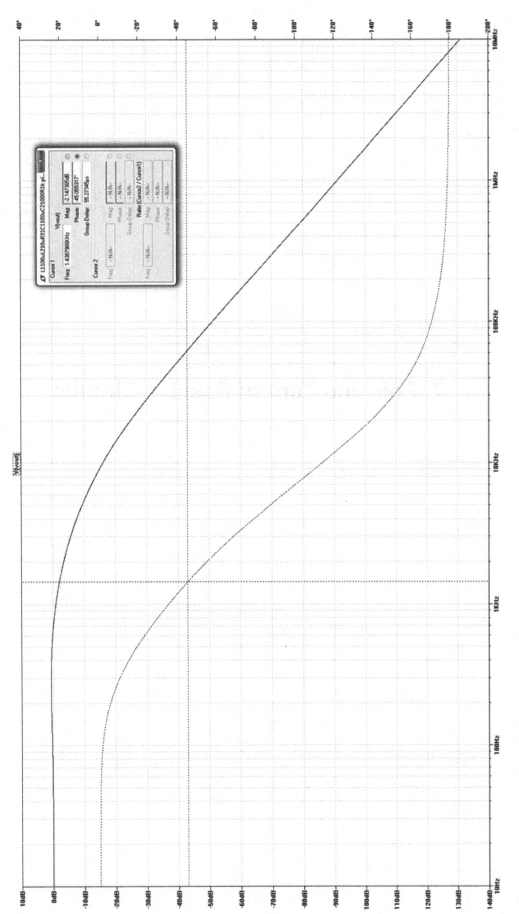

LTSpice simulated results as shown above.

7. Pole and Zero of Passive Circuits

7. Pole and Zero of Passive Circuits

1. Single pole
2. Single pole and zero
3. Double poles
4. Double poles and zero

(Also see Appendix A-4 in Pole and Zero Transfer Function Derivation)

NETWORK	WAVE FORM	$\left(\frac{V_o}{V_i}\right)$ Transfer Function
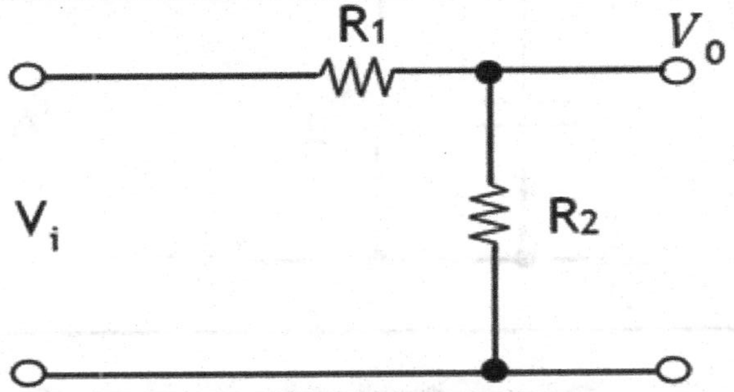	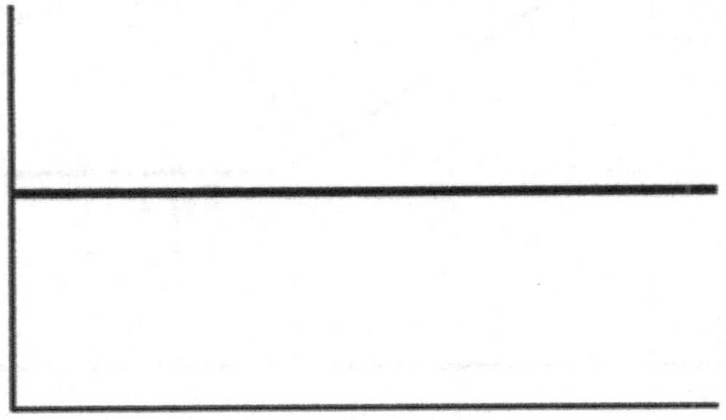	$\dfrac{R_2}{R_1+R_2}$ $\emptyset = 0^0$
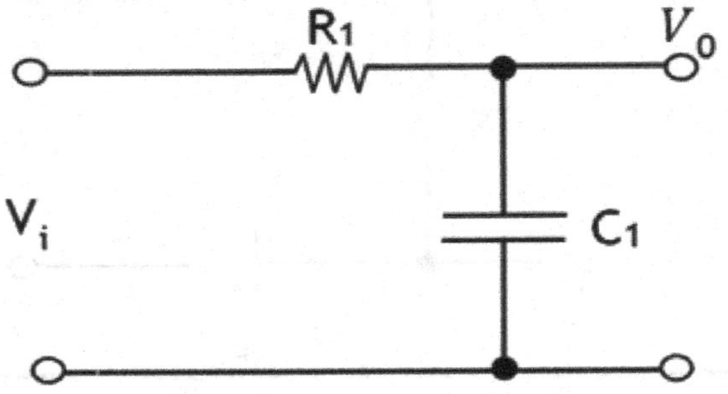	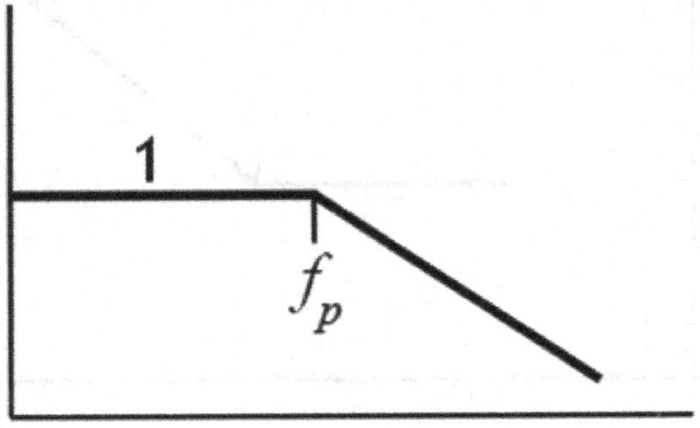	$\dfrac{1}{1+sR_1C_1}$ $\emptyset = -tan^{-1}\left(\dfrac{f}{f_p}\right)$

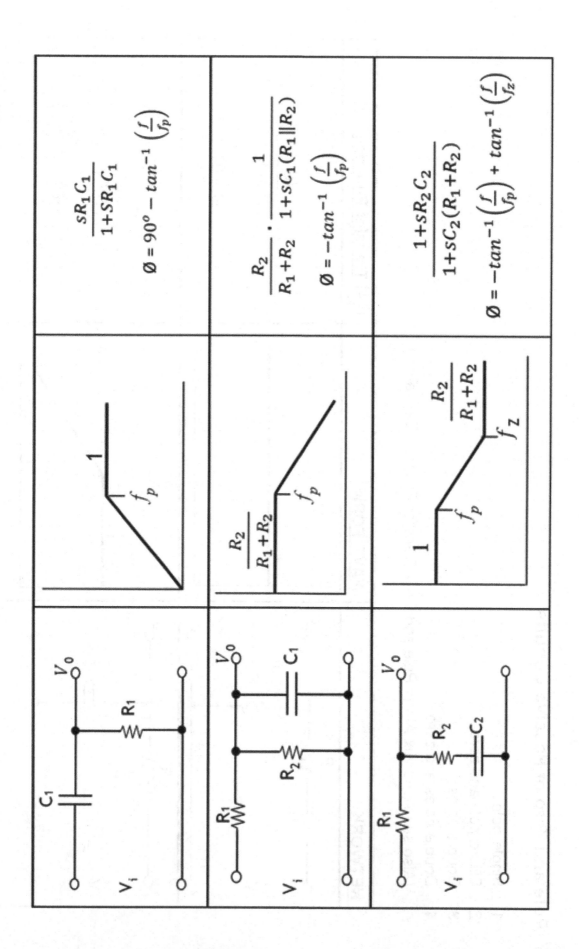

Circuit	Bode Plot	Transfer Function / Phase
V_o, R_1, R_2, C_1, V_i	$\dfrac{R_2}{R_1+R_2}$, f_p	$\dfrac{sC_1R_2}{1+sC_1(R_1+R_2)}$ $\emptyset = 90^o - tan^{-1}\left(\dfrac{f}{f_p}\right)$
V_o, R_1, R_2, C_1, C_2, V_i	$\dfrac{C_1}{C_1+C_2}$, f_p, f_z, $\dfrac{R_2}{R_1+R_2}$ **OR** $\dfrac{C_1}{C_1+C_2}$, f_p, f_z	$\dfrac{C_1}{C_1+C_2} \cdot \dfrac{(sC_2R_2+1)}{sC_1\|C_2(R_1+R_2)+1}$ $\emptyset = -tan^{-1}\left(\dfrac{f}{f_p}\right) + tan^{-1}\left(\dfrac{f}{f_p}\right)$
V_o, R_1, R_2, C_1, V_i	1, f_p, $\dfrac{R_2}{R+R_2}$, f_z	$\dfrac{R_2}{R_1+R_2} \cdot \dfrac{1+sC_1R_1}{1+sC_1(R_1\|R_2)}$ $\emptyset = tan^{-1}\left(\dfrac{f}{f_z}\right) - tan^{-1}\left(\dfrac{f}{f_p}\right)$

$$\frac{R_2}{R_1+R_2} \cdot \frac{1+sC_1R_1}{1+s(C_1+C_2)(R_1\|R_2)}$$

$$\emptyset = tan^{-1}\left(\frac{f}{f_z}\right) - tan^{-1}\left(\frac{f}{f_p}\right)$$

7.1 Single Pole Characteristics:

Single Pole's Gain decreases with increasing in frequency at a rate of -20dB/dec.

Single Pole's Gain = $\dfrac{1}{\left(\frac{s}{\omega_p}+1\right)}$,   where $\omega_p = 2\pi f_p$ and $s = j\omega = j2\pi f$

Gain (dB) = $20\log\left|\dfrac{1}{\left(\frac{s}{\omega_p}+1\right)}\right| = -20\log\left|\left(\dfrac{s}{\omega_p}+1\right)\right| = -20\log\left|\left(\dfrac{j2\pi f}{2\pi f_p}+1\right)\right|$

$\therefore$ Gain = $-20\log\left|j\dfrac{f}{f_p}+1\right|$

But $\left|j\dfrac{f}{f_p}+1\right| = \sqrt{\left(\dfrac{f}{f_p}\right)^2 + 1^2} = \sqrt{\left(\dfrac{f}{f_p}\right)^2 + 1}$ (1)

Plot of Gain in dB:

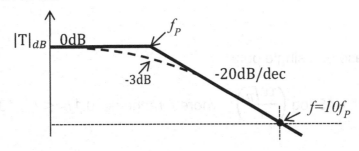

1. For $f \ll f_{P,}$ then Gain $\approx 20\log 1 = 0$dB
2. For $f = f_{P,}$ then Gain $= -20\log\sqrt{2} = -3$ dB at $f = f_P$
3. For $f = 10f_{P,}$ Gain $= -20\log\sqrt{(10)^2+1} \approx -20$dB
4. For $f = 100f_{P,}$ Gain $= -20\log\sqrt{(100)^2+1} \approx -20\log 100 = -40$dB

Conclusion: Gain $\begin{cases} 0\text{dB for any frequency, } f < f_P \\ -\dfrac{20\text{dB}}{\text{decade}} \text{ for frequency, } f > f_P \end{cases}$

<u>**Trick:**</u> To find the value of gain at any frequency, $f > f_P$

$$\boxed{\text{Eq1. Gain (dB) for } f \geq f_P = -20\log\left(\frac{f}{f_p}\right)}$$

$$\boxed{\text{Gain (dB) for } f < f_P = 0\text{dB}}$$

Phase Angle of Single Pole Characteristics:

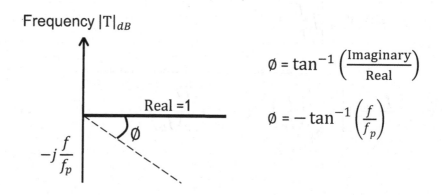

$$\emptyset = \tan^{-1}\left(\frac{\text{Imaginary}}{\text{Real}}\right)$$

$$\emptyset = -\tan^{-1}\left(\frac{f}{f_p}\right)$$

<u>Trick:</u> Phase of a single pole

$$\emptyset_P = -90° + 45°\log\left(\frac{10f_p}{f}\right), \text{ where } f \text{ range is: } 0.1f_P \leq f \leq 10f_P$$

<u>Single Pole Example:</u>

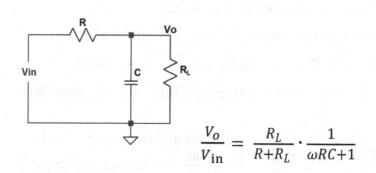

$$\frac{V_o}{V_{in}} = \frac{R_L}{R+R_L} \cdot \frac{1}{\omega RC+1}$$

7.2 Single Pole and Zero Characteristics:

Single Pole's Gain decreases with increasing in frequency at a rate of -20dB/dec.

Single Zero's Gain increases with increasing in frequency at a rate of $+20$dB/dec.

$$\text{Gain (zero)} = \left|\left(\frac{s}{\omega_z} + 1\right)\right| = \sqrt{\left(\frac{f}{f_z}\right)^2 + 1}$$

$$\text{Gain (dB)} = 20\log\sqrt{\left(\frac{f}{f_z}\right)^2 + 1}$$

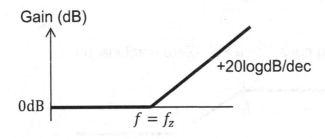

Tricks:

$$\text{Gain(dB)} = -20\log\left(\frac{f}{f_z}\right) \text{ for } f \geq f_z$$

$$\varnothing_z = 90° - 45°\log\left(\frac{10f_z}{f}\right), \text{ where } f \text{ range: } 0.1f_z \leq f \leq 10f_z$$

Phase Angle of Single Zero & Pole Characteristics:

- Maximum Leading phase for a Zero-Pole combination is:

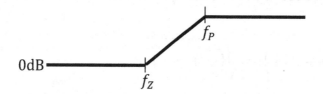

$$\emptyset_{\text{margin}} \text{ Lead } = +45°\log\left(\frac{f_p}{f_z}\right), \qquad \text{Range: } f_Z \leq f_P \leq 100f_Z$$

- Maximum Lagging phase for a Pole-Zero combination is:

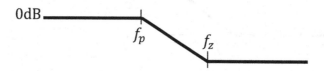

$$\emptyset_{\text{margin}} \text{ Lag } = -45°\log\left(\frac{f_p}{f_z}\right), \qquad \text{Range: } f_P \leq f_Z \leq 100f_P$$

Single Pole and Zero Example:

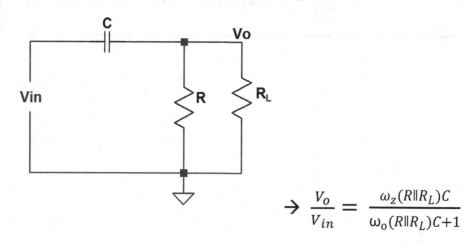

$$\rightarrow \frac{V_o}{V_{in}} = \frac{\omega_z(R\|R_L)C}{\omega_o(R\|R_L)C+1}$$

7.3 Double Pole Characteristics:

Single Pole's Gain decreases with increasing in frequency at a rate of -20dB/dec. Double Poles' Gain may decrease with increasing in frequency at a rate of -40dB/dec. Double poles can be real or complex. Double poles with two real roots of H(s) filter quadratic equation are called real poles. Double poles with complex roots are addressed below.

Complex poles:

H(s) double poles can be represented in the following quadratic equation, which Q is the characteristics of complex double poles and will result the system's resonant frequencies and instability with drastic phase changes as a function of frequency.

$$H(s) = \frac{1}{\left[\dfrac{s}{\omega_0}\right]^2 + \dfrac{1}{Q} \cdot \dfrac{s}{\omega_0} + 1}$$

$$\text{Gain (magnitude)} = 20\log(H(s)) = -20\log\left|\left[\frac{s}{\omega_0}\right]^2 + \frac{1}{Q} \cdot \frac{s}{\omega_0} + 1\right|$$

$$\text{Gain(dB)} = -20\log\left|\left[\left[j\frac{2\pi f}{2\pi f_0}\right]^2 + \frac{1}{Q} \cdot j\frac{2\pi f}{2\pi f_0} + 1\right]\right| = -20\log\left|1 - \left[\frac{f}{f_0}\right]^2 + j\frac{f}{Qf_0}\right|$$

$$\text{Gain(dB)} = -20\log\sqrt{\left(1 - \left[\frac{f}{f_0}\right]^2\right)^2 + \left(\frac{f}{Qf_0}\right)^2}$$

For $f = f_o$

$$\text{Gain(dB)} = -20\log\sqrt{(0)^2 + \left(\frac{1}{Q}\right)^2} = -20\log\left(\frac{1}{Q}\right)$$

$$\boxed{\text{Gain(dB) (at } f=f_o) = +20\log(Q)}$$

Hence for Q > 1 we have peaking for $f \gg f_o$, where

$$\text{Gain(dB)(at } f \gg f_o) \approx -20\log\sqrt{\left(\left[\frac{f}{f_0}\right]^2\right)^2 + \left(\frac{f}{Qf_0}\right)^2} \approx -20\log\left[\frac{f}{f_0}\right]^2 = -40\text{Log}\left(\frac{f}{f_0}\right)$$

$$\boxed{\text{Gain(dB)(at } f \gg f_0) = -40\log\left(\frac{f}{f_0}\right)}$$

Phase of Double Complex Pole Characteristics:

$$\emptyset = -\tan^{-1}\left(\frac{\text{Imaginary}}{\text{Real}}\right) = -\tan^{-1}\left(\frac{\frac{f}{Qf_0}}{1-\left(\frac{f}{f_0}\right)^2}\right) = \tan^{-1}\left(\frac{1}{Q}\cdot\frac{\frac{f}{f_0}}{1-\left(\frac{f}{f_0}\right)^2}\right)$$

For $f = f_o$ $Q = -\tan^{-1}\left(\frac{1}{Q}\right)\left(\frac{1}{0}\right) = -\tan^{-1}\infty = -90°$

For $f \gg f_o$ $Q = -\tan^{-1}\left(\frac{1}{Q}\cdot\frac{f_o}{f}\right)$

Double Pole Example:

$$\rightarrow \quad \frac{V_o}{V_{in}} = \frac{K}{\left(\frac{s}{\omega_0}\right)^2 + \frac{1}{Q}\cdot\frac{s}{\omega_0} + 1}$$

7.4 Double Poles and Zero Characteristics

Double poles and zero H(s) are very similar to Double Pole Characteristics above. The differences are that it introduces a zero in the capacitor C_2 and R_d damping circuit, where $C_2 > 3C$. The math represetation is shown below.

$$H(s) = \frac{\dfrac{1}{s/\omega_z + 1}}{\left(\dfrac{s}{\omega_0}\right)^2 + \dfrac{1}{Q} \cdot \dfrac{s}{\omega_0} + 1}$$

Double Poles and Zero Example:

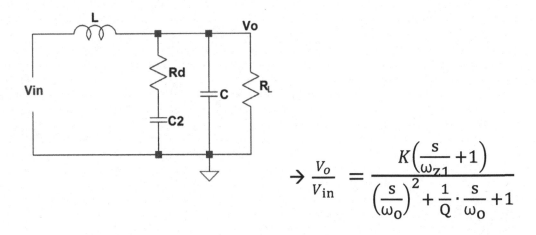

$$\rightarrow \frac{V_o}{V_{in}} = \frac{K\left(\dfrac{s}{\omega_{z1}} + 1\right)}{\left(\dfrac{s}{\omega_0}\right)^2 + \dfrac{1}{Q} \cdot \dfrac{s}{\omega_0} + 1}$$

8. Poles and Zero of Active Circuits

8. Poles and Zero of Active Circuits

ERROR AMPLIFIER CIRCUIT	WAVEFORM	$\left(\dfrac{\hat{V}_c}{\hat{V}_x}\right)$ TRANSFER FUNCTION
(op-amp circuit with R_1, R_2, $\hat{V}_x$, $\hat{V}_c$, V_{REF})	(magnitude plot of $\frac{\hat{V}_c}{\hat{V}_x}$ at level R_F/R_1 vs f)	$\dfrac{\hat{V}_c}{\hat{V}_x} = -\dfrac{R_F}{R_1}$ $\emptyset = 0^o$
(op-amp circuit with R_1, C_1, V_x, V_c, V_{REF})	(plot of $\frac{\hat{V}_c}{\hat{V}_x}$ with slope from $\frac{1}{sC_1R_1}$)	$\dfrac{\hat{V}_c}{\hat{V}_x} = -\dfrac{1}{sC_1R_1}$ $\emptyset = -\tan^{-1}\dfrac{\omega C_1 R_1}{0} = -\tan^{-1}\infty = -90^0$
(op-amp circuit with R_1, R_F, C_s, $\hat{V}_x$, $\hat{V}_c$, V_{REF})	(plot of $\frac{\hat{V}_c}{\hat{V}_x}$ with breakpoints R_F/R_1, $\frac{1}{2\pi C_s R_1}$, $\frac{R_F}{\frac{1}{sC_s}} = sC_s R_F$)	$\dfrac{\hat{V}_c}{\hat{V}_x} = -\dfrac{R_F}{R_1}(sC_s R_1 + 1)$ $\emptyset = \tan^{-1}\omega C_s R_1$

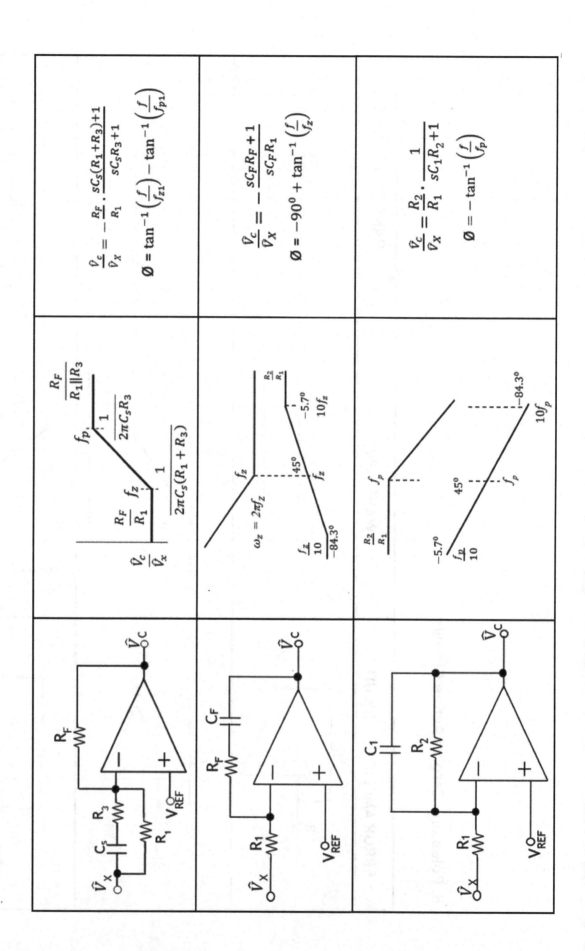

$$\frac{\hat{v}_c}{\hat{v}_x} = -\frac{R_F}{R_1} \cdot \frac{sC_s(R_1+R_3)+1}{sC_sR_3+1}$$

$$\emptyset = \tan^{-1}\left(\frac{f}{f_{z1}}\right) - \tan^{-1}\left(\frac{f}{f_{p1}}\right)$$

$$\frac{\hat{v}_c}{\hat{v}_x} = -\frac{sC_FR_F+1}{sC_FR_1}$$

$$\emptyset = -90^0 + \tan^{-1}\left(\frac{f}{f_z}\right)$$

$$\frac{\hat{v}_c}{\hat{v}_x} = \frac{R_2}{R_1} \cdot \frac{1}{sC_1R_2+1}$$

$$\emptyset = -\tan^{-1}\left(\frac{f}{f_p}\right)$$

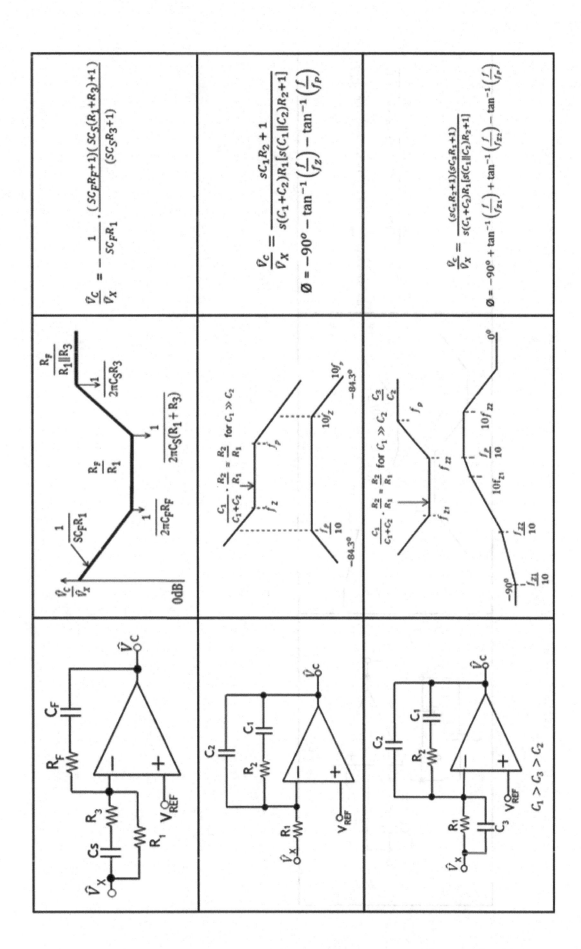

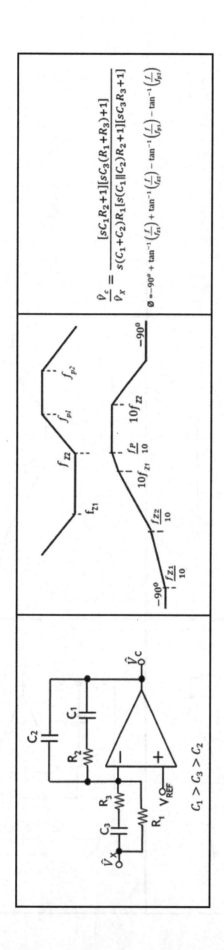

$$\frac{\hat{v}_c}{\hat{v}_x} = \frac{[sC_1R_2+1][sC_3(R_1+R_3)+1]}{s(C_1+C_2)R_1[s(C_1\|C_2)R_2+1][sC_3R_3+1]}$$

$$\emptyset = -90^\circ + \tan^{-1}\left(\frac{f}{f_{z1}}\right) + \tan^{-1}\left(\frac{f}{f_{z2}}\right) - \tan^{-1}\left(\frac{f}{f_{p1}}\right) - \tan^{-1}\left(\frac{f}{f_{p2}}\right)$$

$C_1 > C_3 > C_2$

9. Comparing Open vs. Closed Loop Feedback Characteristics and Their Effects on Phase Margin and Queing

9.1 Single Pole Open Loop System:

$$A = \frac{A(0)}{\left(\dfrac{s}{\omega_a} + 1\right)}$$

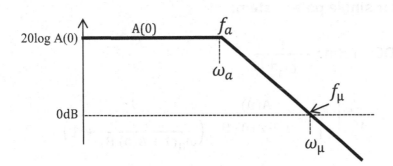

For unity gain *(20log(A) = 0dB → 1 at ω_μ)*, then $\dfrac{s}{\omega_a}$ >> 1 (because A(0) is open loop gain, >>1)

$$\therefore \qquad A \approx \frac{A(0)}{s/\omega_a} = \frac{A(0)\cdot\omega_a}{s}$$

Set
$$A = 1 = \frac{A(0)\cdot\omega_a}{s_\mu} = \frac{A(0)\cdot\omega_a}{\omega_\mu}$$

Therefore

$$\boxed{\omega_\mu \approx A(0)\cdot\omega_a \quad \text{or} \quad f_\mu = A(0)\cdot f_a}$$

which is defined at gain A(0), bandwidth $= f_a$, and the product (A(0)· f_a) is a constant.

9.2 Single Pole Closed Loop System:

$$\left.\frac{V_o}{V_{in}}\right|_{C.L} = \frac{A}{1+AB} \quad , \qquad \text{where } A = \frac{A(0)}{\left(\frac{s}{\omega_a}+1\right)}$$

$$\left.\frac{V_o}{V_{in}}\right|_{C.L} = \frac{\frac{A(0)}{s/\omega_a + 1}}{1 + \frac{A(0)}{\left(s/\omega_a+1\right)}\cdot B} = \frac{A(0)}{\left(\frac{s}{\omega_a} + 1 + A(0)\cdot B\right)}$$

General form is for single pole system:

$$\left.\frac{V_o}{V_{in}}\right|_{C.L} = DC\_\ Term \cdot \frac{1}{\left(s/\omega_a+1\right)}$$

$$\text{i.e.} \ \therefore \ \left.\frac{V_o}{V_{in}}\right|_{C.L} = \frac{A(0)}{1 + A(0)\cdot B} \cdot \frac{1}{\left(\frac{s}{\omega_a(1 + A(0)\cdot B)} + 1\right)}$$

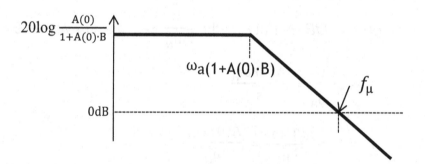

At unity gain f_μ , we set:

$$\left.\frac{V_o}{V_{in}}\right|_{C.L} = 0dB \ \rightarrow \ \text{Gain} = 1$$

and therefore:
$$\frac{s}{\omega_a(1+A(0)\cdot B)} \gg 1$$

$$\therefore \qquad \left.\frac{V_o}{V_{in}}\right|_{C.L} = 1 = \frac{A(0)}{\cancel{(1+A(0)\cdot B)}} \cdot \frac{\cancel{\omega_a(1+A(0)\cdot B)}}{s_\mu}$$

$$\omega_\mu = A(0)\omega_a$$

$$f_\mu = A(0)f_a$$

$\therefore$ which product relationship is same as in open loop.

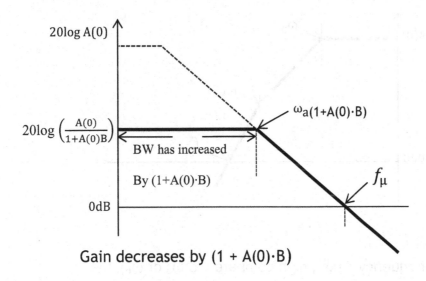

Gain decreases by (1 + A(0)·B)

9.3 Two Pole Open Loop System:

$$A = \frac{A(0)}{\left(\dfrac{s}{\omega_a} + 1\right)\left(\dfrac{s}{\omega_b} + 1\right)}$$

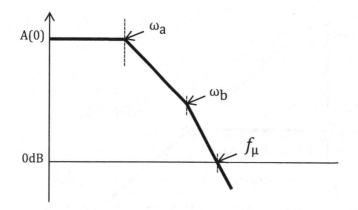

Note: at High frequency f (i.e., high compared to ω_a or ω_b), then

$$A \approx \frac{A(0)}{\left(\dfrac{s}{\omega_a}\right)\left(\dfrac{s}{\omega_b}\right)} = \frac{A(0)\omega_a\omega_b}{s_\mu^2} = \frac{A(0)\omega_a\omega_b}{\omega_\mu^2}$$

Therefore;

$$\boxed{\begin{aligned} \omega_\mu &= \sqrt{A(0)\omega_a\omega_b} \\ f_\mu &= \sqrt{A(0)f_a f_b} \end{aligned}}$$

9.4 Two Pole Closed Loop System:

$$\left.\frac{V_o}{V_{in}}\right|_{C.L} = \frac{A}{1+A\cdot B} \ , \text{ where } A = \frac{A(0)}{\left(\frac{s}{\omega_a}+1\right)\left(\frac{s}{\omega_b}+1\right)}$$

$$\left.\frac{V_o}{V_{in}}\right|_{C.L} = \frac{A}{1+A\cdot B} = \frac{\frac{A(0)}{\left(\frac{s}{\omega_a}+1\right)\left(\frac{s}{\omega_b}+1\right)}}{1+\frac{A(0)\cdot B}{\left(\frac{s}{\omega_a}+1\right)\left(\frac{s}{\omega_b}+1\right)}} = \frac{A(0)}{\left(\frac{s}{\omega_a}+1\right)\left(\frac{s}{\omega_b}+1\right)+A(0)\cdot B} = \frac{A(0)}{\frac{s^2}{\omega_a\omega_b}+s\left(\frac{1}{\omega_a}+\frac{1}{\omega_b}\right)+(1+A(0)\cdot B)}$$

$$\left.\frac{V_o}{V_{in}}\right|_{C.L} = \left[\frac{A(0)}{(1+A(0)\cdot B)}\right]\left[\frac{1}{\frac{s^2}{\omega_a\omega_b(1+A(0)\cdot B)}+\frac{s}{(1+A(0)\cdot B)}\left(\frac{1}{\omega_a}+\frac{1}{\omega_b}\right)+1}\right]$$

We next want to evaluate how the effects of a two pole open loop system affects the Q of a closed loop system as the second pole moves from a higher frequency towards the open loop bandwidth.

Evaluating the Q of the closed loop response at the band width, fc of the of the open loop gain.

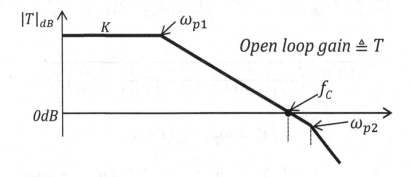

For closed loop response, we need $\left|\frac{T}{1+T}\right|_{dB}$

Note: $\frac{T}{1+T} = \begin{cases} 1, \ for \ T > 1 \\ T, \ for \ T < 1 \end{cases}$

Plot of $\left|\frac{T}{1+T}\right|_{dB}$ vs. frequency:

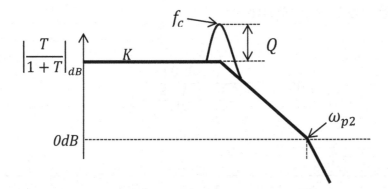

We need to find the amplitude of Q. Since T is a two pole roll off, we have

$$T = \frac{K}{\left(\frac{s}{\omega_{p1}} + 1\right)\left(\frac{s}{\omega_{p2}} + 1\right)}$$

Substitute in $\dfrac{T}{1+T}$:

$$\frac{T}{1+T} = \frac{\dfrac{K}{\left(\frac{s}{\omega_{p1}} + 1\right)\left(\frac{s}{\omega_{p2}} + 1\right)}}{1 + \dfrac{K}{\left(\frac{s}{\omega_{p1}} + 1\right)\left(\frac{s}{\omega_{p2}} + 1\right)}}$$

$$\frac{T}{1+T} = \frac{K}{\left(\frac{s}{\omega_{p1}} + 1\right)\left(\frac{s}{\omega_{p2}} + 1\right) + K} = \frac{K}{\left(\frac{s^2}{\omega_{p1}\omega_{p2}} + s\left(\frac{1}{\omega_{p1}} + \frac{1}{\omega_{p2}}\right) + 1 + K\right)}$$

$$= \frac{K}{1+K} \cdot \frac{1}{\left(\frac{s^2}{\omega_{p1}\omega_{p2}(1+K)} + \frac{s}{(1+K)} \cdot \left(\frac{1}{\omega_{p1}} + \frac{1}{\omega_{p2}}\right) + 1\right)}$$

$$\therefore \omega_o^2 = \omega_{p1}\omega_{p2}(1+K)$$

$$\frac{1}{Q\omega_o} = \frac{1}{1+K}\left(\frac{1}{\omega_{p1}} + \frac{1}{\omega_{p2}}\right) = \frac{1}{1+K}\left(\frac{\omega_{p1} + \omega_{p2}}{\omega_{p1}\omega_{p2}}\right)$$

$$Q = \frac{1}{\omega_o} \cdot (1+K)\left(\frac{\omega_{p1}\omega_{p2}}{\omega_{p1} + \omega_{p2}}\right)$$

$$Q = \frac{1}{\sqrt{\omega_{p1}\omega_{p2}(1+K)}} \cdot (1+K)\left(\frac{\omega_{p1}\omega_{p2}}{\omega_{p1} + \omega_{p2}}\right)$$

$$Q = \frac{\sqrt{\omega_{p1}\,\omega_{p2}(1+K)}}{\omega_{p1} + \omega_{p2}} = \sqrt{\frac{\omega_{p1}\| \omega_{p2}}{\omega_{p1} + \omega_{p2}}}\,(1+K)$$

For $\omega_{p2} \gg \omega_{p1}$, then $\omega_{p1}\|\omega_{p2} \approx \omega_{p1}$ and $\omega_{p1} + \omega_{p2} \approx \omega_{p2}$

Then,
$$\boxed{Q = \sqrt{\frac{\omega_{p1}}{\omega_{p2}}}\,(1 + K)}$$

This is *Q* of the closed loop response due to an open loop two pole system,

ω_{p1} and ω_{p2}.

We can evaluate whether we have real or imaginary roots by Evaluating Q.

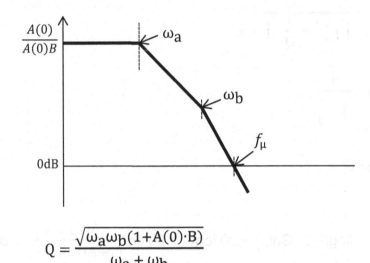

$$Q = \frac{\sqrt{\omega_a\omega_b(1+A(0)\cdot B)}}{\omega_a + \omega_b}$$

Let's look at ω_a , ω_b two pole system and their placement with respect to ω_μ or f_μ.

1.   If $\omega_a = \omega_b = \omega_\mu$, then $Q = \dfrac{\sqrt{\omega_\mu^{\,2}(1+A(0)\cdot B)}}{2\omega_\mu} = \dfrac{\sqrt{(1+A(0)\cdot B)}}{2}$

   If $A(0) \cdot B \gg 1$, then $Q \approx \dfrac{\sqrt{A(0)\cdot B}}{2}$

   Hence if $\dfrac{\sqrt{A(0)\cdot B}}{2} > \dfrac{1}{2}$ or $\sqrt{A(0) \cdot B} > 1$, which it generates a very large Q.

2. If $\omega_a = \omega_b$, we have complex roots.

3. If $\omega_b \gg \omega_a$, then $Q = \dfrac{\sqrt{\omega_a\omega_b(1+A(0)\cdot B)}}{\omega_a+\omega_b} \approx \sqrt{\dfrac{\omega_a\omega_b}{\omega_b^2}(1 + A(0) \cdot B)}$

$$Q = \sqrt{\frac{\omega_a}{\omega_b}(1 + A(0) \cdot B)}$$

$$\left.\frac{V_o}{V_{in}}\right|_{C.L} = DC\_Gain \cdot \cfrac{1}{\left(\dfrac{s}{\omega_0}\right)^2 + \dfrac{1}{Q} \cdot \dfrac{s}{\omega_0} + 1}$$

$$s = j\omega_o$$

At s = $j\omega_0$, we have

$$\left.\frac{V_o}{V_{in}}\right|_{C.L} = DC\_Gain \cdot \cfrac{1}{[\,j\,]^2 + \dfrac{1}{Q} \cdot j + 1}$$

$$\left.\frac{V_o}{V_{in}}\right|_{C.L} = DC\_Gain \cdot \cfrac{1}{j \cdot \dfrac{1}{Q}}$$

$$\left.\frac{V_o}{V_{in}}\right|_{C.L} = DC\_Gain \cdot Q$$

$$20\log\left(\left.\frac{V_o}{V_{in}}\right|_{C.L}\right) = 20\log(DC\_Gain) + 20\log Q = 20\log\frac{A(0)}{(1+A(0)\cdot B)} + 20\log Q$$

4. If $\omega_b = \omega_\mu$ or $f_b = f_\mu$ ($\omega_b \gg \omega_a$), then

$$Q = \frac{\sqrt{\omega_a \omega_b (1+A(0)\cdot B)}}{\omega_a + \omega_b} \approx \sqrt{\frac{\omega_a \omega_b}{\omega_b^2}(1+A(0)\cdot B)} = \sqrt{\frac{\omega_a}{\omega_b}(1 + A(0)\cdot B)}$$

At unity gain f_μ, we set: $\left.\dfrac{V_o}{V_{in}}\right|_{C.L} = 0dB \;\rightarrow\; 1$

$$\left.\frac{V_o}{V_{in}}\right|_{C.L.} = \left[\frac{A(0)}{(1+A(0)\cdot B)}\right]\left[\cfrac{1}{\dfrac{s^2}{\omega_a\omega_b(1+A(0)\cdot B)} + \dfrac{s}{(1+A(0)\cdot B)}\left(\dfrac{1}{\omega_a} + \dfrac{1}{\omega_b}\right) + 1}\right]$$

Since $f_\mu = A(0)f_a f_b$, then

$$Q = \sqrt{\frac{\omega_a}{\omega_b}(1 + A(0) \cdot B)} = \sqrt{\frac{2\pi f_a}{2\pi f_b}(1 + A(0) \cdot B)} = \sqrt{\frac{2\pi f_a}{2\pi f_\mu}(1 + A(0) \cdot B)}$$

$$Q = \sqrt{\frac{2\pi f_a}{2\pi A(0) f_a f_b}(1 + A(0) \cdot B)} = \sqrt{\frac{2\pi f_a}{A(0) f_b}(1 + A(0) \cdot B)}$$

$$\boxed{Q = 1}$$

In other words, when the second pole f_b is near the unity gain bandwidth, ω_μ or f_μ , the two pole system will generate a Q in a closed loop system as shown in the above math calculation, which the Q will create an unexpected instability under transient conditions in a closed loop system.

The **open loop gain can not** predict the closed loop response when the 2nd pole is at or near 0dB (<2 octave before or above 0dB) at unity gain frequency. The open loop gain can only predict the closed loop gain if the feedback system has a single pole roll off passed unity gain, and any second poles are located at least 2 octave away from 0dB after the unity gain bandwidth.

We will demonstrate these phenomena via LTSpice in class.

General Form for Two Pole System:

$$\frac{V_o}{V_{in}}\bigg|_{C.L} = DC\_Gain \cdot \frac{1}{\left[\frac{s}{\omega_o}\right]^2 + \frac{1}{Q} \cdot \frac{s}{\omega_o} + 1}$$

$$\omega_0^2 = \omega_a\omega_b(1 + A(0) \cdot B), \quad DC\_Gain = \frac{A(0)}{1 + A(0) \cdot B}$$

$$\omega_0 = \sqrt{\omega_a\omega_b(1 + A(0) \cdot B)}$$

Note: $\dfrac{1}{\omega_a} + \dfrac{1}{\omega_b} = \dfrac{\omega_a + \omega_b}{\omega_a\omega_b}$

$$\frac{V_o}{V_{in}}\bigg|_{C.L} = DC\_Gain \cdot \frac{1}{\left[\frac{s}{\omega_o}\right]^2 + \frac{s(\omega_a + \omega_b)}{\omega_a\omega_b(1 + A(0) \cdot B)} + 1}$$

Note that: $\omega_0 = \omega_a\omega_b(1 + A(0) \cdot B)$

$$\therefore \quad \frac{V_o}{V_{in}}\bigg|_{C.L} = DC\_Gain \cdot \frac{1}{\left[\frac{s}{\omega_o}\right]^2 + \left[\frac{s}{\omega_o}\right]\left[\frac{\omega_a + \omega_b}{\omega_o}\right] + 1}$$

$$\therefore \quad \frac{1}{Q} = \frac{\omega_a + \omega_b}{\omega_o} \quad \rightarrow \quad Q = \frac{\omega_o}{\omega_a + \omega_b} = \frac{\sqrt{\omega_a\omega_b(1 + A(0) \cdot B)}}{\omega_a + \omega_b}$$

10. Roots of General Quadratic Equation: $y = ax^2 + bx + c$

$$y_{1,2} = \frac{-b \pm \sqrt{b^2 - 4ac}}{2a} \quad \text{and} \quad \frac{1}{s_{1,2}} = \frac{1}{\left[\frac{s}{\omega_0}\right]^2 + \frac{1}{Q}\cdot\frac{s}{\omega_0} + 1} \quad \rightarrow \quad \frac{1}{as^2 + bs + c}$$

$$a = \frac{1}{\omega_0{}^2} \; , \quad b = \frac{1}{Q\cdot\omega_0} \quad \text{and} \quad c = 1$$

$$s_{1,2} = \frac{-\frac{1}{Q\omega_0} \pm \sqrt{(\frac{1}{Q\omega_0})^2 - 4\cdot\frac{1}{\omega_0{}^2}(1)}}{2(\frac{1}{\omega_0{}^2})} = \frac{-\frac{1}{Q\omega_0} \pm \frac{1}{Q\omega_0}\sqrt{1 - 4Q^2}}{2(\frac{1}{\omega_0{}^2})}$$

$$\boxed{s_{1,2} = \frac{\omega_0}{2Q}\left[-1 \pm \sqrt{1 - 4Q^2}\right]}$$

Note for a real roots of $(s_{1,2})$, $\sqrt{1 - 4Q^2} \geq 1$

Let's find the boundary condition when $\sqrt{1 - 4Q^2} = 0$:

$$\sqrt{1 - 4Q^2} = 0 \quad \rightarrow \quad 1 - 4Q^2 = 0$$

$$\boxed{Q = \sqrt{\frac{1}{4}} = \frac{1}{2}}$$

For $Q \leq \frac{1}{2}$, we have real roots.

For $Q \geq \frac{1}{2}$, we have complex roots.

Since $\sqrt{1 - 4Q^2}$, $Q > \frac{1}{2}$ is an imaginary number (i.e., $j(\sqrt{1 - 4Q^2})$).

Since $Q = \frac{\sqrt{\omega_a\omega_b(1 + A(0)\cdot B)}}{\omega_a + \omega_b}$

11. AC Model of Buck Circuit Including Closed Loop Design

In order to design any converter, we must close the loop to keep either the output voltage constant or the output current constant. To close the loop, we need to model (AC model) the various converter stages (called "power stage"). Our objective is to derive a linear model for the non-linear power stage by considering small signal perturbations.

AC Model:

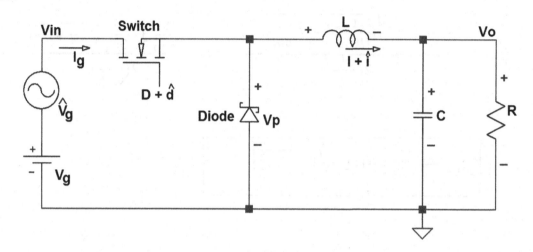

For AC Voltage Small Signal Model

$$d^* \cdot V_g^* = (V_g + \hat{V}_g)(D + \hat{d})$$

$$d^* \cdot V_g^* = D \cdot V_g + D \cdot \hat{V}_g + V_g \cdot \hat{d} + \hat{V}_g \cdot \hat{d})$$

where $D \cdot V_g$ is DC term; $D \cdot \hat{V}_g + V_g \cdot \hat{d}$ are AC terms; $\hat{V}_g \cdot \hat{d}$ is non-linear term.

Set both DC and non-linear terms to zero.

$$d^* \cdot V_g^* = D \cdot \hat{V}_g + V_g \cdot \hat{d}$$

For AC Current Small Signal Model

$$d^* \cdot i^* = (I + \hat{\imath})(D + \hat{d})$$

$$d^* \cdot i^* = D \cdot I + D \cdot \hat{\imath} + I \cdot \hat{d} + \hat{\imath} \cdot \hat{d}$$

where $D \cdot I$ is DC term; $D \cdot \hat{\imath} + I \cdot \hat{d}$ are AC terms; $\hat{\imath} \cdot \hat{d}$ is non-linear term.

Set both DC and non-linear terms to zero.

$$d^* \cdot i^* = D \cdot \hat{\imath} + I \cdot \hat{d}$$

DC Model:

$$V_p = D \cdot V_g \qquad \text{set} \qquad \hat{V}_g = 0, \ \hat{V}_o = 0$$

$$I_g = I \cdot D \qquad\qquad\qquad \hat{d} = 0$$

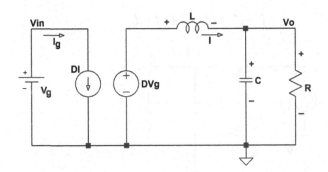

For DC Model: L → Short   and   C → Open

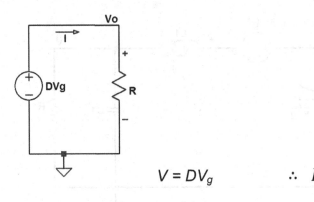

$$V = DV_g \qquad\qquad \therefore \ D = \frac{V}{V_g}$$

$$I = \frac{V}{R}$$

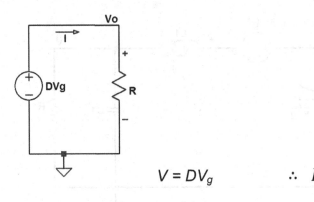

$$I \cdot D = \frac{V}{R} \cdot D$$

$$P_{in} = V_g \cdot \frac{V}{R} \cdot D \qquad \text{but } V_g = \frac{V}{D}$$

$$\therefore \ P_{in} = \frac{V}{D} \cdot \frac{V}{R} \cdot D = \frac{V^2}{R}$$

For AC Model:　　　　set DC and non-linear terms to zero

$$d^* \cdot V_g^* = D \cdot \hat{V}_g + V_g \cdot \hat{d}$$

$$d^* \cdot i^* = D \cdot \hat{\imath} + I \cdot \hat{d}$$

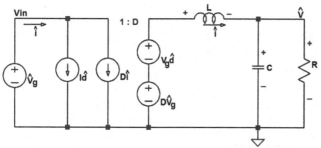

where $D\hat{\imath}$ and $D\hat{V}_g$ are ideal Transformer in AC Terms

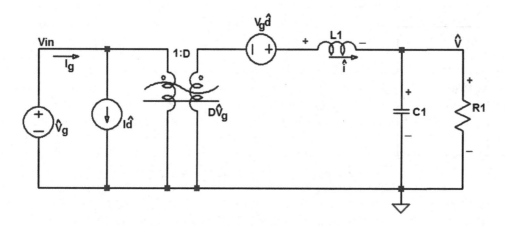

Therefore from the above equations we have :

$$\boxed{T = \frac{V_g}{V_m} H(s) \cdot A(s)}$$

We can now find $GV_g\big|_{C.L} = FV_g$, and $Z_{OL}\big|_{C.L}$.

Hence $FV_g = \dfrac{GV_g}{1 + T}$,　　where GV_g is the output to input transfer

Function $T \triangleq$ Loop Gain

$$Z_{OL}\big|_{C.L} = \frac{Z_{O.L}}{1 + T} \text{ , Where } Z_{O.L} \text{ is the open loop output impedance.}$$

So we must Find GV_g and $Z_{O.L}$:

For GV_g we open the loop and set $\hat{d} = 0$ so that we can find

$$GV_g = \left. \frac{\hat{V}}{\hat{V}_g} \right|_{\text{Open\_Loop (O.L)}}$$

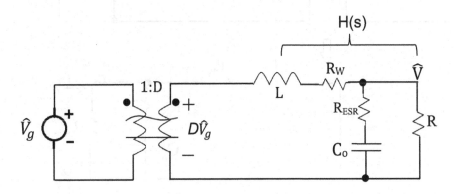

$$\boxed{ GV_g = \left. \frac{\hat{V}}{\hat{V}_g} \right|_{O.L} = D \cdot H(s) }$$

Next we find Z_{OL} by setting both $\hat{V}_g$ and $\hat{d}$ to zero $\hat{V}_g = \hat{d} = 0$ and letting R = R_{nom} (The Nominal Load)

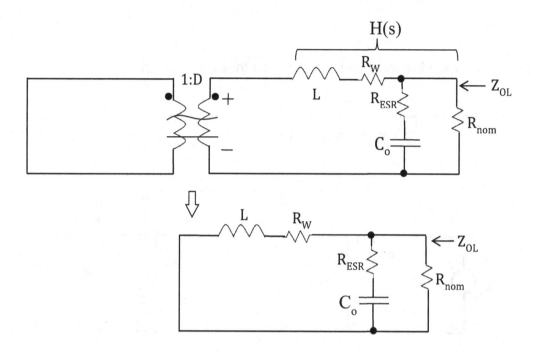

Let's draw the Z_{OL} plot from Low to High frequency, where Z_{OL} is looking back impedance at open_loop output node.

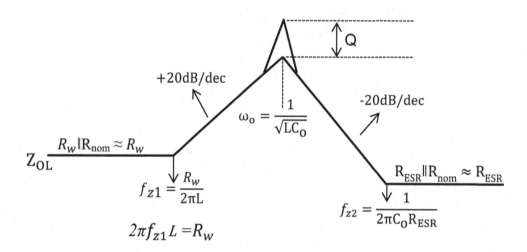

Note: $Z_{OL} = \dfrac{R_W(sC_oR_{ESR} + 1)}{sC_o(R_{ESR} + R_{nom}) + 1}$: At high frequency between ω_o and f_{z2}

12. Buck Converter Design Example

Find output-to-input closed loop, C.L, Transfer function, $\left.\dfrac{\hat{V}(s)}{\hat{V}_g(s)}\right|_{C.L}$. Also find output-to-load

Transfer function closed, $\left.\dfrac{\hat{V}(s)}{\hat{I}_{LOAD}(s)}\right|_{C.L}$.

In order to find $\left.\dfrac{\hat{V}(s)}{\hat{V}_g(s)}\right|_{C.L}$ and $\left.\dfrac{\hat{V}(s)}{\hat{I}_{LOAD}(s)}\right|_{C.L}$, we must first find the loop gain, T .

Given information:

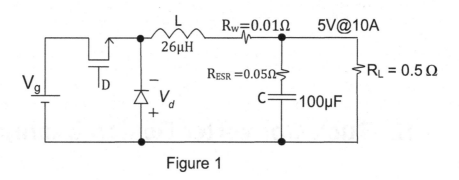

Figure 1

For a given input voltage range: $15V \leq V_g \leq 30V$

Let's look at the Q due to the Low pass filter, H(s). Then draw the AC model of Figure 1 by adding the feedback elements (i.e. Modulator and Error Amplifier Gain)

Solving for Q:

$$R_o = \sqrt{\frac{L}{C}} = \sqrt{\frac{26\mu H}{100uF}} = 0.51,$$

$$Q_L = \frac{R_w + R_L}{R_o} = \frac{0.01 + 0.5}{0.51} = 1 \quad \text{and} \quad Q_c = \frac{R_o}{(R_w \| R_L) + R_{ESR}} = \frac{0.51}{(0.01\|0.5) + 0.05} = 8.5$$

$Q = Q_L \| Q_c = 0.9$

Draw AC Model:

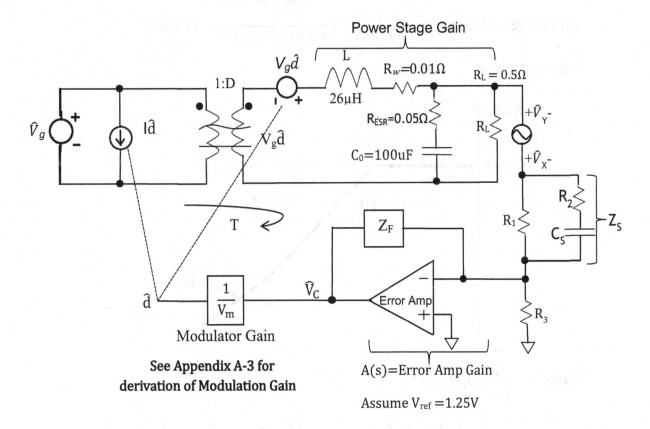

See Appendix A-3 for
derivation of Modulation Gain

$A(s)$ = Error Amp Gain

Assume V_{ref} = 1.25V

Note: $T = |A(s)| = \dfrac{Z_F}{Z_S}$

$$T \triangleq \frac{\hat{V}_Y}{\hat{V}_X} = \left(\frac{\hat{V}_Y}{\hat{d}}\right) \cdot \left(\frac{\hat{d}}{\hat{V}_c}\right) \cdot \left(\frac{\hat{V}_c}{\hat{V}_X}\right)$$

$$T = (Power\_Stage\_Gain) \cdot (Modulation\_Gain) \cdot (Error\_Amp\_Gain)$$

$$T = V_g H(s) \cdot \left(\frac{1}{V_m}\right) \cdot A(s)$$

$$\therefore \quad T = \frac{V_g}{V_m} \cdot H(s) \cdot (A(s)$$

Since Q = 0.9 we have two poles at $f_o = \dfrac{1}{2\pi\sqrt{26\mu H(100\mu F)}} = 3.1kHz$,

and an ESR Zero at $f_z = \dfrac{1}{2\pi C(ESR)} = \dfrac{1}{2\pi(100\mu F)(0.05\Omega)} = 31.8kHz$

Plot of $|H(s)|_{dB}$

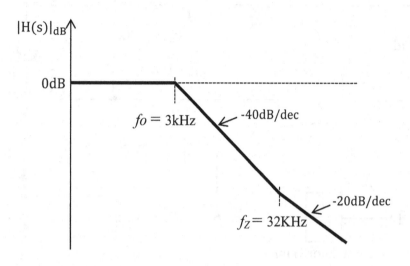

Note: $T = \dfrac{V_g}{V_m} \cdot H(s) \cdot A(s)$

Since $\hat{V}(s) = \left[1 + \dfrac{R_1}{R_2}\right]\left(\dfrac{T}{1+T}\right)\hat{V}_{ref} + \underbrace{\left(\dfrac{GV_g}{1+T}\right)\hat{V}_g + \dfrac{Z_{out} + \hat{I}_{LOAD}(s)}{1+T}}_{Disturbances}$

where $GV_g = DH(s)$ and Z_{OUT} = open_loop_output impedance. We can see that we want T as large as possible, also when we are at the bandwidth f_c, we would like phase margin $\emptyset_m > 45^0$, so we avoid the Q due to the closed loop response (i.e., $\dfrac{T}{1+T}$).

Hence we want $\dfrac{K}{s}$ behaves as an integrator with gain of K.

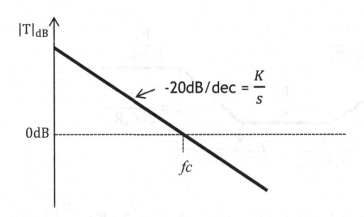

Hence: $\varnothing_m > 90^0$

So we have $T = \left(\dfrac{V_g}{V_m}\right) \cdot H(s) \cdot (A(s))$ and we want $T = \dfrac{K}{s} \;\rightarrow\; A(s) = \dfrac{K'}{s \cdot H(s)}$

Then $T = \dfrac{V_g}{V_m} \cdot H(s) \cdot \left(\dfrac{K'}{s \cdot H(s)}\right) = \dfrac{V_g}{V_m} \cdot \dfrac{K'}{s} = \dfrac{K}{s}$, where $K = \dfrac{V_g}{V_m} \cdot K'$

Since H(s) have two poles and one zero we need A(s) to be as follows;

A(s) Design Characteristics:

     1. Two zeros to cancel the two poles of H(s)

     2. One pole to cancel the ESR zero of H(s)

     3. An integrator for high loop gain T

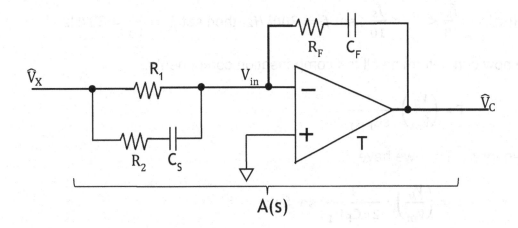

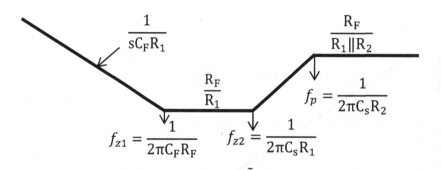

So we see from A(s) that $K' = \dfrac{1}{C_F R_1}$ $\rightarrow$ $K = \dfrac{V_g}{V_m} \cdot K' = \dfrac{V_g}{V_m} \cdot \dfrac{1}{C_F R_1}$

Next is to determine where to place f_{Z1} and f_{Z2}.

Since H(s) have two poles at f_o = 3kHz. We want one zero before f_o and one zero after f_o to avoid peaking.

Let f_{Z1} = 2kHz and f_{Z2} = 4kHz, also $f_p = f_{ZESR}$

We must now decide on our bandwidth, f_c, since f_o = 3kHz and we know that;

$$F \triangleq \frac{GV_g}{1+T} = \left| \frac{\hat{V}(s)}{\hat{V}_g(s)} \right|_{C.L} = \frac{DH(s)}{1+T}$$

This tells us that when T = 1 (0dB), then $F \triangleq \left| \dfrac{\hat{V}(s)}{\hat{V}_g(s)} \right|_{C.L}$ will not help attenuate. The input signal, $\hat{V}_g$.

If we set f_C = 30kHz, then at f_C we gain -46dB attenuation of the input signal from

$|DH(s)|_{dB}$. $\dfrac{f_s}{20} \le f_c \le \dfrac{f_s}{10}$, set f_s = 300kHz, then set $f_c = \dfrac{f_s}{10} = 30kHz$

We now can determine all the compensation components.

$$T = \left(\frac{V_g}{V_m} \right) \cdot \frac{1}{sC_F R_1}$$

Then for f_C, T = 1 we have

$$f_C = \left(\frac{V_g}{V_m} \right) \cdot \frac{1}{2\pi C_F R_1} \quad \text{so that:}$$

$$\boxed{C_F = \frac{V_g}{V_m} \cdot \frac{1}{2\pi f_c R_1}}$$

R_1 is part of the voltage divider to obtain V_{OUT}

$$V = V_{OUT} = \left[1 + \frac{R_1}{R_3}\right] \cdot V_{ref}$$

Since V= 5V and V_{ref} = 1.25V

Then $\frac{R_1}{R_3} = 3$ Let R_3 = 10kΩ, then R_1 = 30kΩ

Note: T is a function of V_g, so we pick V_{g_nom} = 22V for f_C = 30kHz, also given is V_m = 2V

$$C_F = \frac{V_g}{V_m} \cdot \frac{1}{2\pi f_c R_1} = \frac{22V}{2V} \cdot \frac{1}{2\pi(30kHz)(30k\Omega)} = 1.95nF$$

Use

$$\boxed{C_F = 2nF = 2000pF}$$

$$f_{Z1} = \frac{1}{2\pi C_F R_F} = 2kHz \quad \rightarrow \quad R_F = \frac{1}{2\pi(2kHz)(0.2nF)} = 40k\Omega$$

$$\boxed{R_F = 40k\Omega}$$

$$f_{Z2} = \frac{1}{2\pi C_S R_1} = 4kHz \quad \rightarrow \quad C_S = \frac{1}{2\pi(4kHz)(30k\Omega)} = 1.3nF$$

$$\boxed{C_S = 1.3nF}$$

$$f_p = \frac{1}{2\pi C_S R_2} = 32kHz \quad \rightarrow \quad R_2 = \frac{1}{2\pi(32kHz)(1.3nF)} = 3.8k\Omega$$

$$\boxed{R_2 = 3.8k\Omega}$$

Compensation:

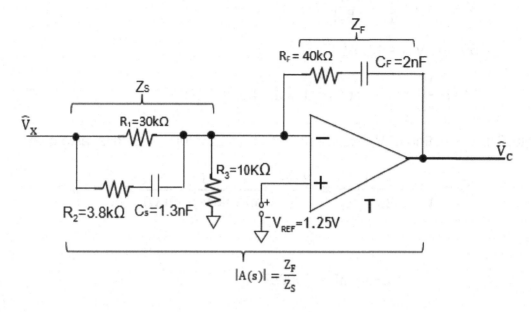

$$|A(s)| = \frac{Z_F}{Z_S}$$

Next, we want to plot $F = \dfrac{GV_g}{1+T} = \dfrac{GH(s)}{1+T} = \left|\dfrac{\hat{V}(s)}{\hat{V}_g(s)}\right|_{C.L}$

$$D = \frac{V}{V_{g\_nom}} = \frac{5V}{22V} = 0.227$$

So $20\log D = 20\log(0.227) = -12.9\text{dB}$

Hence we need to plot $|F|_{dB} = 20\log(|F|) = \left|\dfrac{DH(s)}{1+T}\right|_{dB}$

$$|F|_{dB} = 20\log D + 20\log H(s) + 20\log\left(\frac{1}{1+T}\right)$$

So we plot each term in dB, then add each graph

OPEN LOOP OUTPUT IMPEDANCE (Z$_{OL}$)

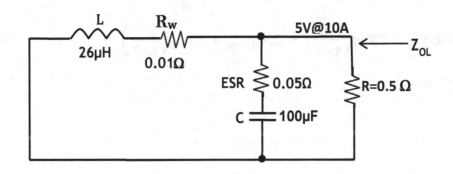

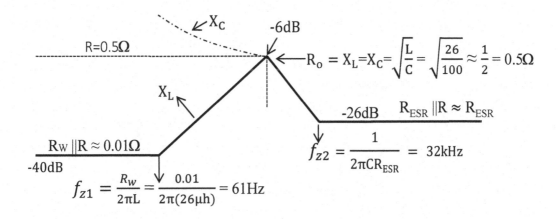

$$f_{z1} = \frac{R_w}{2\pi L} = \frac{0.01}{2\pi(26\mu h)} = 61Hz$$

Note: $Q = 1$ (This is the same Q as we calculated before $Q = Q_L \| Q_C$)

Let $1\Omega = 0dB$

Then $\qquad 20\log\left(\frac{R_w \| R}{1\Omega}\right) = 20\log\left(\frac{0.01}{1\Omega}\right) = -40dB$

When $\qquad R_o = X_L = X_C = \sqrt{\frac{L}{C}} = 0.5\Omega \qquad \rightarrow \qquad 20\log\left(\frac{0.5}{1\Omega}\right) = -6dB$

At high frequency $\ Z_o = ESR = 0.05\Omega \qquad \rightarrow \quad 20\log\left(\frac{0.05\Omega}{1\Omega}\right) = -26dB$

CLOSE LOOP OUTPUT IMPEDANCE (Z$_{OL}$)

Now we can determine the closed loop Gain by adding $20\log\left(\frac{1}{1+T}\right)$ to $20\log Z_{OL}$,

$$Z_0 = \frac{Z_{OL}}{1+T} \text{ (This is the closed loop output impedance)}$$

$$20\log Z_0 = 20\log\left(\frac{Z_{OL}}{1+T}\right) = 20\log Z_{OL} + 20\log\left(\frac{1}{1+T}\right)$$

Note: 20log T is:

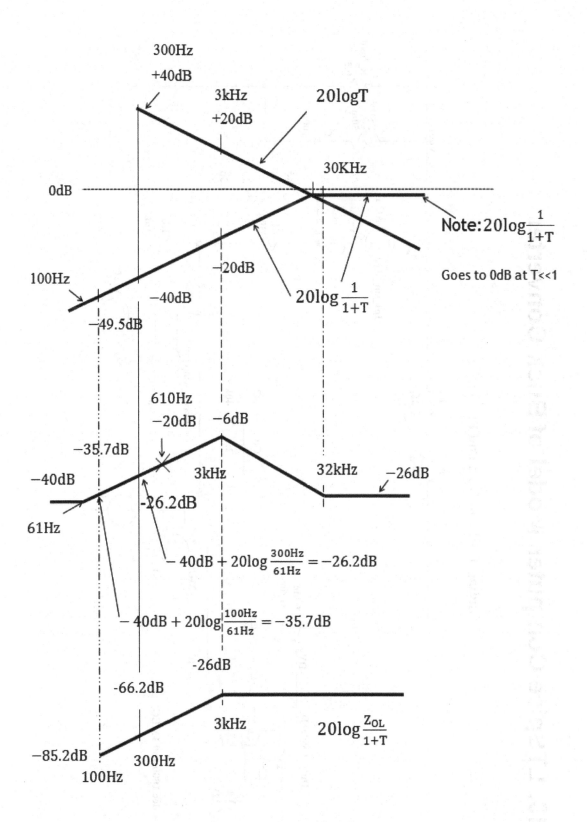

13. LTSpice Computer Model of Buck Converter

Finding T = Loop Gain and Phase:

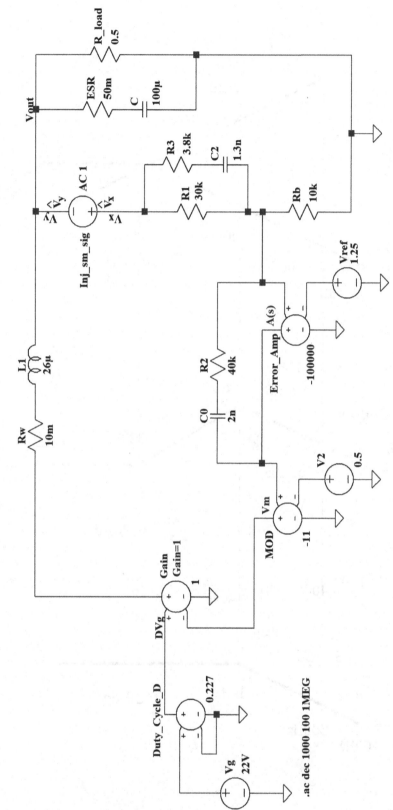

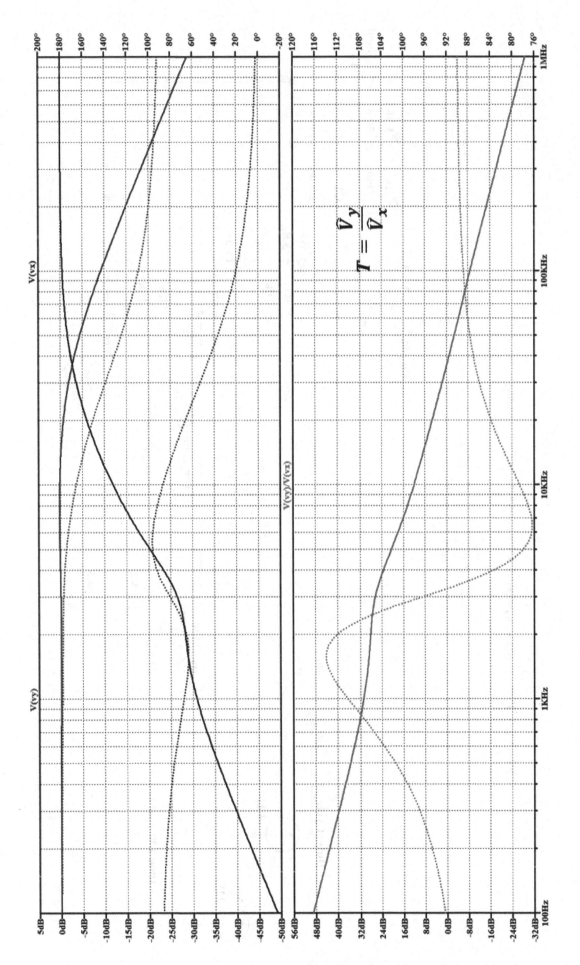

$$T = \frac{\hat{V}_y}{\hat{V}_x}$$

Finding T = Loop Gain and Phase with Q Damping:

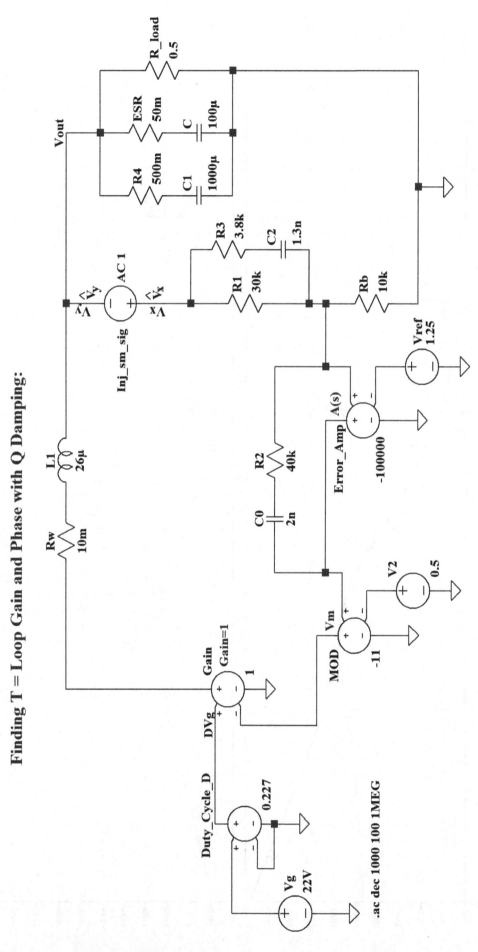

De-Qued T

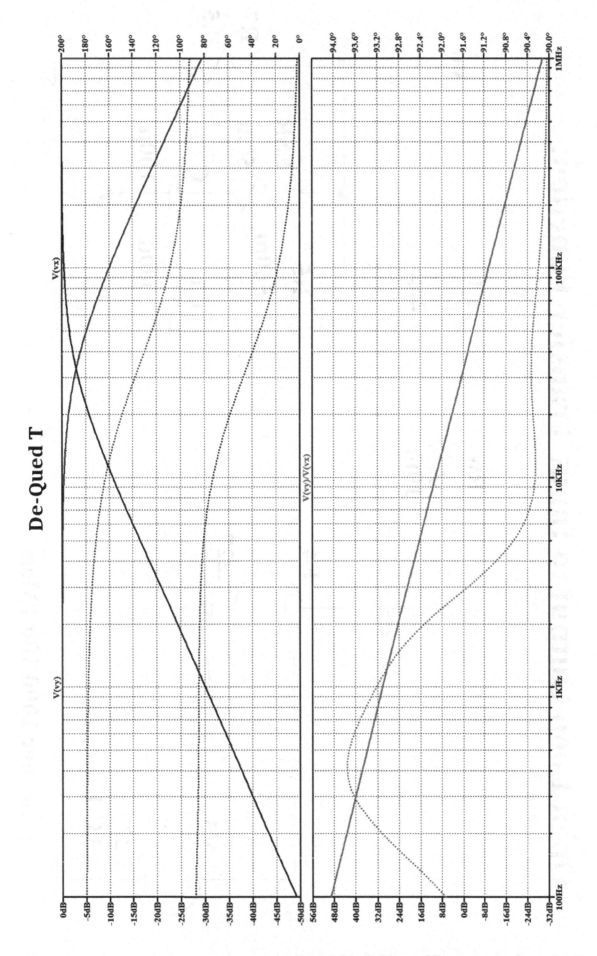

Open Loop Output to Input Transfer Function:

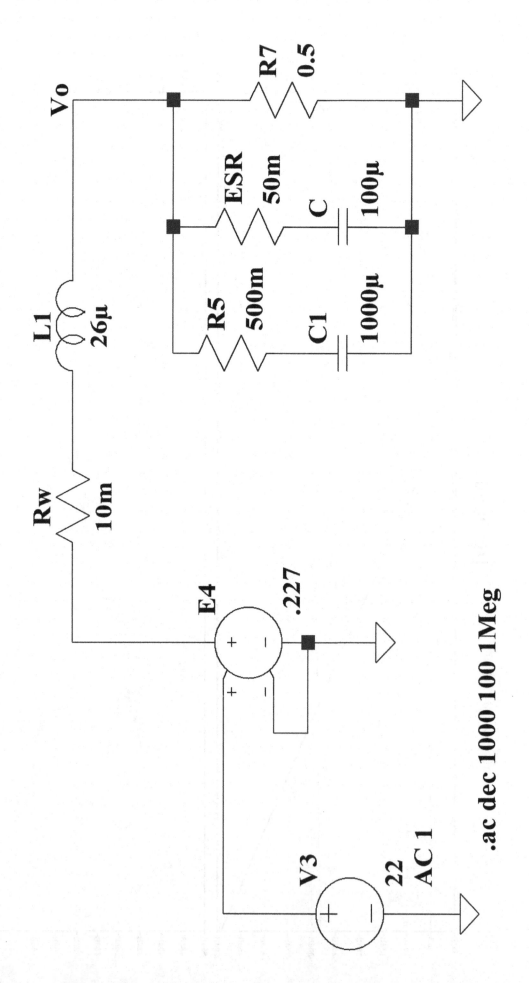

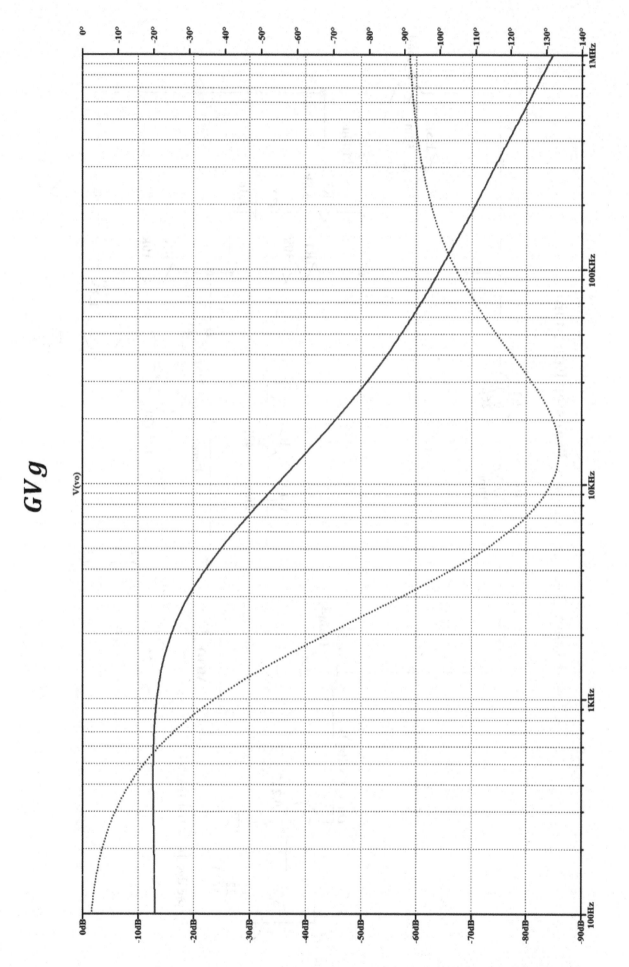

GV g

Closed Loop Output to Input Transfer Function:

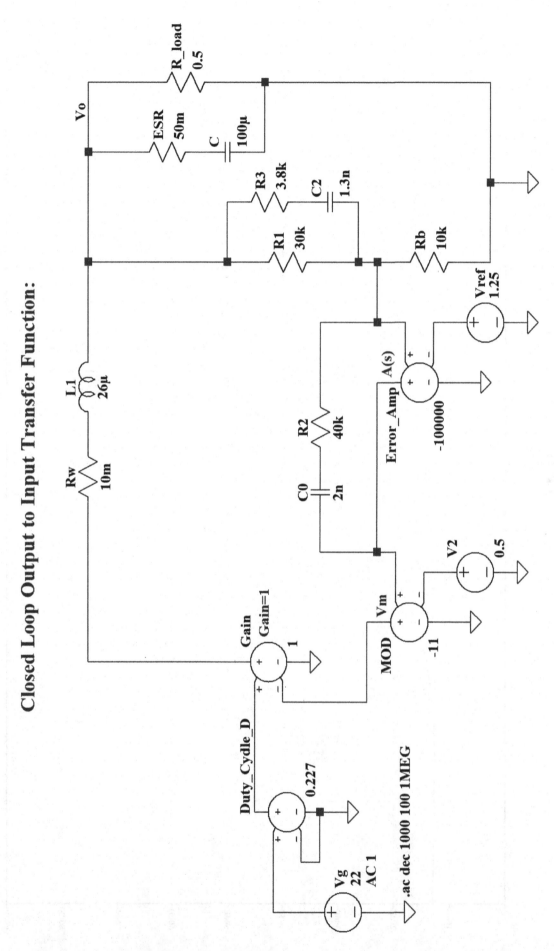

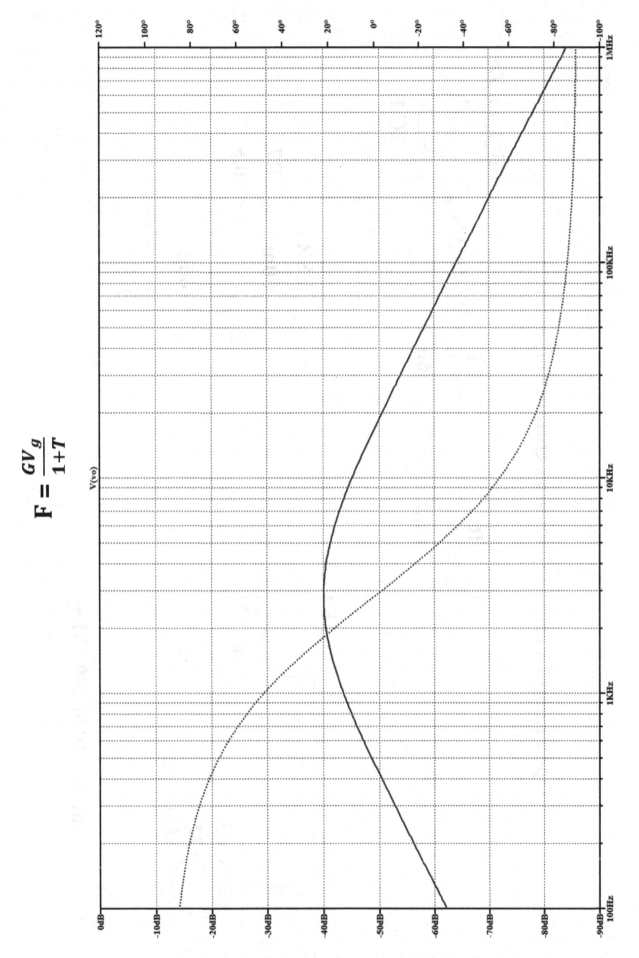

$$\mathbf{F} = \frac{GV_g}{1+T}$$

Open Loop Current Injection at Output Impedance:

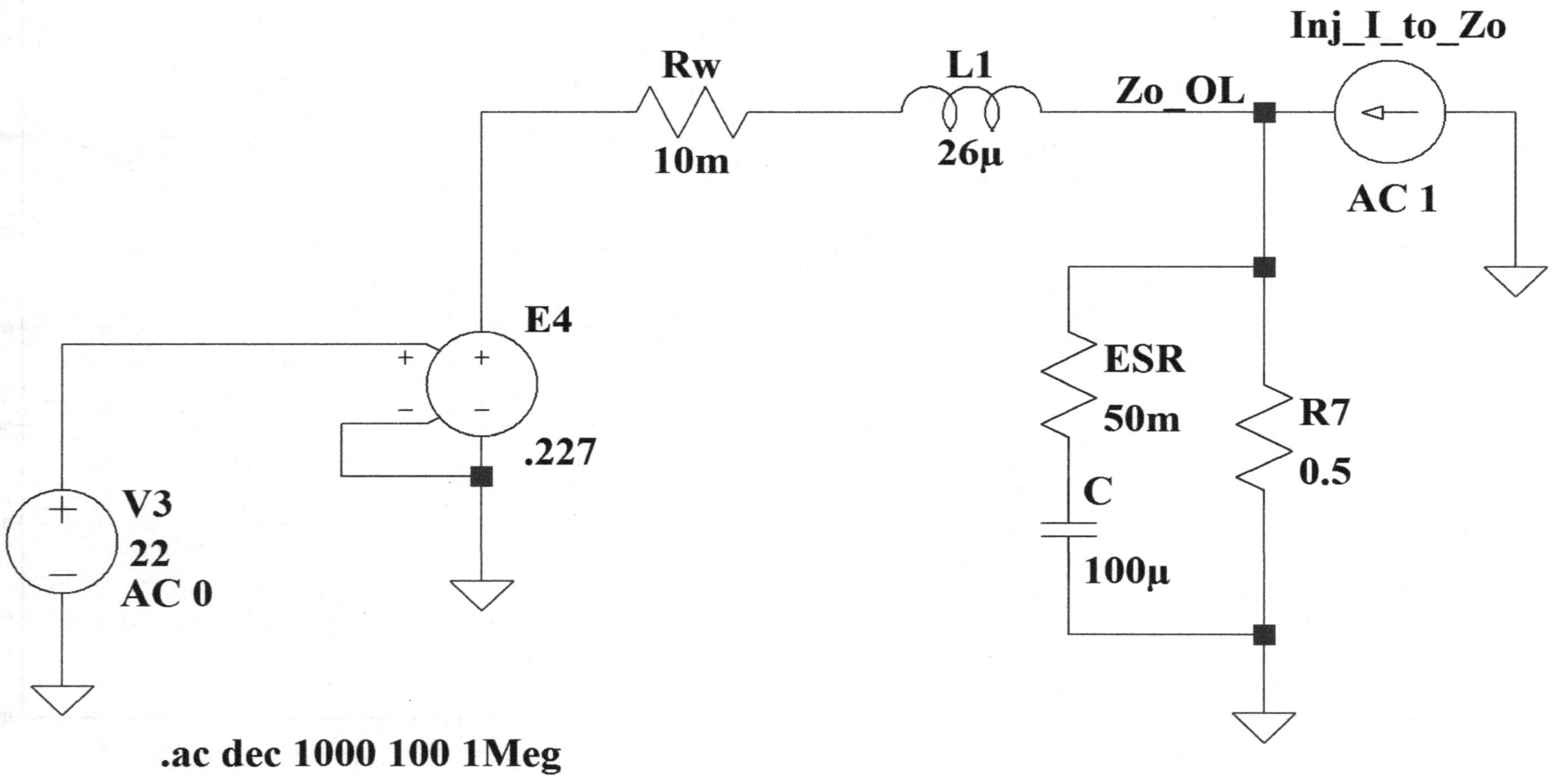

Inj_I_to_Zo

AC 1

Zo_OL

R7
0.5

L1
26μ

ESR
50m

C
100μ

Rw
10m

E4

.227

V3
22
AC 0

.ac dec 1000 100 1Meg

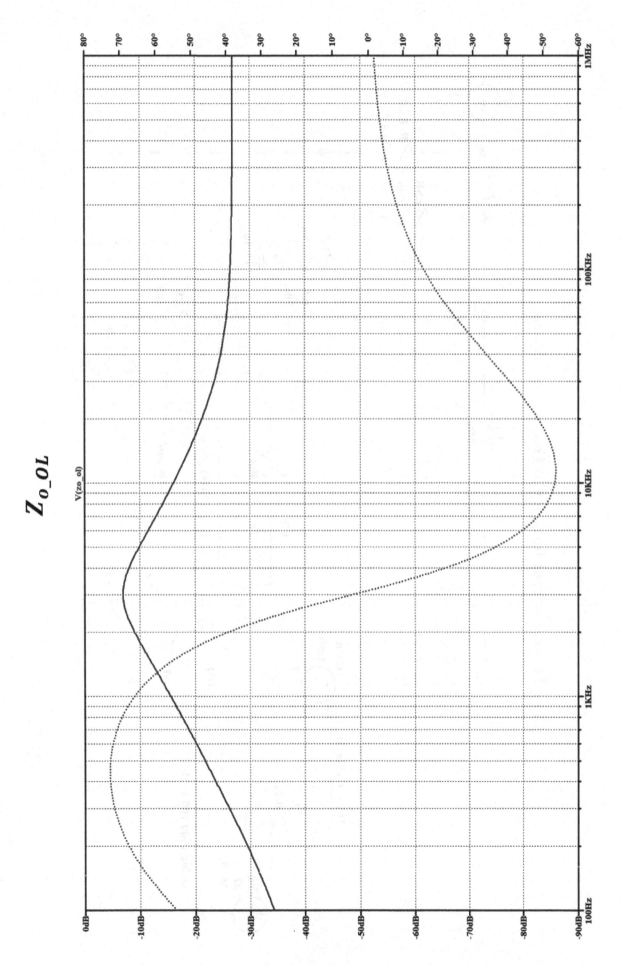

Z_{o_OL}

Closed Loop Current Injection at Output Impedance:

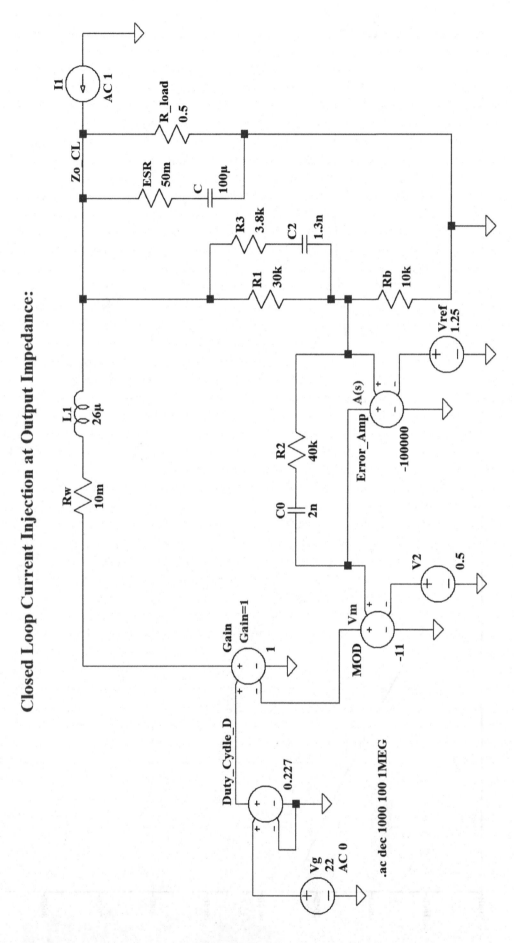

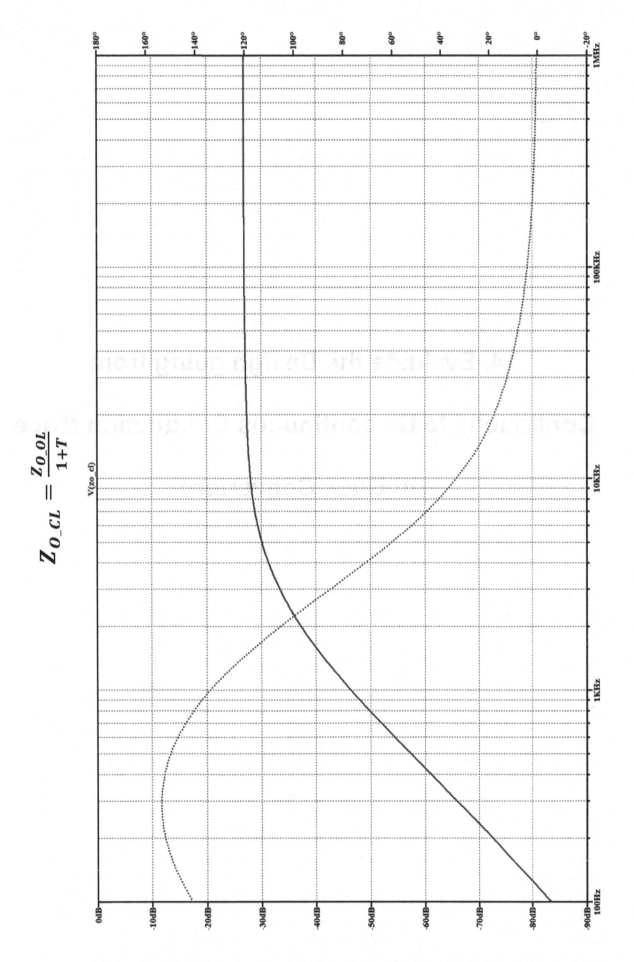

$$Z_{O\_CL} = \frac{Z_{O\_OL}}{1+T}$$

V(zo_cl)

14. Evaluate the Design going from Continuous to Discontinuous Conduction Mode in Buck Converter

14.1 Continuous Conduction Mode (CCM) Buck Converter

Design CCM(Continuous Conduction Mode) Buck converter, then change the design for DCM (Discontinuous CM).

Design Requirements:

V_{ref} = 1.25V	V_o = 2.5V	I_{o_min} = 1A	I_{o_max} = 10A
$V_{ripple} \leq 50mV_{p-p}$	V_m = 2V	f_s =300kHz	$5V \leq V_g \leq 25V$

Derived Requirements:

$V_g = V_{in\_nom} = 15V$ $\qquad I_{nominal} = \dfrac{I_{o\_max}}{2} = 5A$ $\qquad R_{nominal} = 2.5V/5A = 0.5\Omega$

$I_{ripple} = \dfrac{I_{o\_max}}{10} = 1A_{p-p}$ (assume I_{ripple} is $\dfrac{1}{10}$ of I_{o_max})

$T_s = \dfrac{1}{f_s} = \dfrac{1}{300kHz} = 3.33\mu sec$

Design For CCM

$$GV_g = G_{og} = \frac{\hat{V}}{\hat{V}_g} = \frac{1 + \dfrac{s}{\omega_p}}{\left(\dfrac{s}{\omega_O}\right)^2 + \dfrac{1}{Q} \cdot \dfrac{s}{\omega_O} + 1}$$

(note: Gog = Gain of output to generator voltages)

During D' period

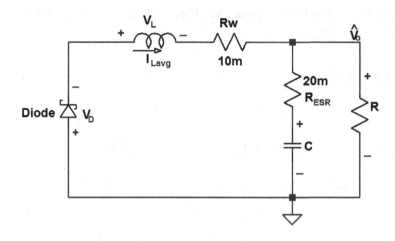

$$V_L = V_O + V_D = L \frac{di}{dt}\bigg|_{dt=D'T_s}$$

$$L = \frac{(V_O + V_D)}{\Delta i} \cdot D' \cdot Ts$$

USE

$$\Delta I = \frac{I_{o\_max}}{10} = \frac{10A}{10} = 1A_{p\text{-}p}$$

$$L \geq \frac{2.5V + 0.7V}{1A_{p-p}} \cdot D' \cdot T_S$$

Since we want L_{max}, use $D'_{max} \rightarrow D_{min}$

$$D_{min} = \frac{V_O}{V_{in\_max}} = \frac{2.5}{25} = 0.1$$

$$D'_{max} = 1 - D_{min} = 0.9$$

$$L_{max} \geq \frac{3.2V}{1A_{p-p}} \cdot \frac{0.9}{300kHz} = 9.6\ \mu H$$

Hence use $\boxed{L = 10\ \mu H}$

Select ESR Capacitor Options:

Now we see that $\text{ESR} \approx \frac{\Delta V}{\Delta i} = \frac{50mV_{p-p}}{1A_{p-p}} = 50m\Omega$

Hence use ESR ≤ 20mΩ, which it exceeds the design requirement.

Check for ESR at cold temperature

Select ESR = 20mΩ, C = 470μF

(Note: The capacitor selection should be first based on the specified desired ESR values over operating temperature range. ESR in electrolytic or tantalum capacitors tend to increase at colder temperature, as shown in the figure below. Consideration of capacitor value should be secondary. Larger capacitor value will increase the current spikes due to rate of change in $C\frac{dv}{dt}$. So larger value in selected capacitor is not necessary better option. Designer needs to consider the application design requirements. However, ESR and Capacitance product do not change over temperature.)

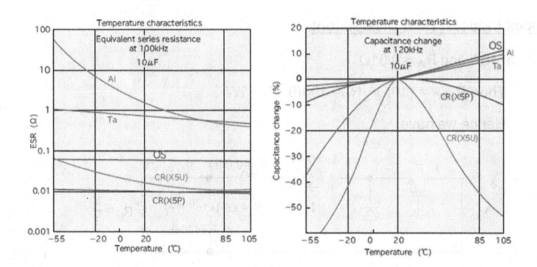

Open Circuit Buck Converter using LTSpice Simulation is (assume η = 90%):

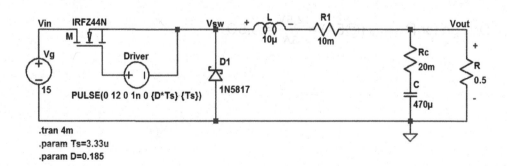

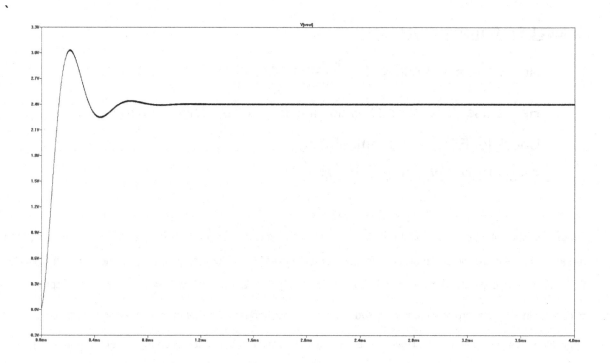

Check Inductor Heat Dissipation:

Note at 10A and $R_W = 0.01\Omega$

Then $P_{DRW} = (10A)^2 \cdot R_W = 100(0.01) = 1W$

Hence we have:

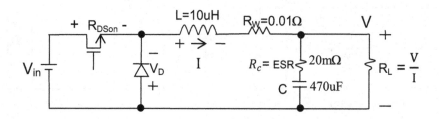

R_L @ 1A = 2.5Ω $0.25 \le R_L \le 2.5\Omega$

R_L @ 10A = 0.25Ω $10A \ge I \ge 1A$

Design for $R_{nominal}$ (@5A) = $\frac{2.5V}{5A}$ = 0.5Ω

Find Q of LCR Network:

$$R_O = \sqrt{\frac{L}{C}} = \sqrt{\frac{1}{47}}$$

$$Q_L = \frac{R_W + R}{R_O} = \frac{0.01 + 2.5}{0.146} = 17.2$$

$$Q_C = \frac{R_O}{(R_W \| R) + R_C} = \frac{0.146}{0.01\|2.5 + 0.02} = 4.87$$

$$\boxed{Q = Q_L \| Q_C = 17.2 \| 4.87 = 3.8}$$

Since Q is large, we must lower the Q by either increasing C or adding a De-Queing circuit across C = 470µF.

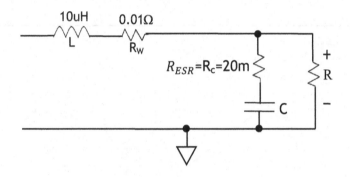

1. Let C = 3·(470uF) ≈ 1500uF, R=2.5Ω, R_c=20mΩ

    Then $R_O = \sqrt{\frac{L}{C}} = \sqrt{\frac{10uH}{1500uF}} = \sqrt{\frac{1}{150}}$

    Then $Q_L = \frac{R_W + R}{R_O} = \sqrt{150}\,(2.51) = 30.7$

    $Q_C = \frac{R_O}{(R_W \| R) + R_C} = \frac{1}{\sqrt{150}(0.03)} = 2.72$

∴ $Q = Q_L \| Q_C = 2.72 \| 30.7 = 2.5$

2. Let's increase C further to reduce Q, where C = 6x450µF= 2700µF.

    Then $Q_L = \sqrt{270}\,(2.51) = 41.24$

    $Q_C = \frac{1}{\sqrt{270}(0.03)} = 2.03$

∴ $Q = Q_L \| Q_C = 2.03 \| 41.24 = 1.933$

3. Let's increase C further, where C = 15x470µF = 7050µF.

Then $Q_L = \sqrt{705}\,(2.51) = 66.7$

$$Q_C = \frac{1}{\sqrt{705}\,(0.03)} = 1.26$$

$\therefore\ Q = Q_L \| Q_C = 54.4 \| 1.26 = 1.23$ (still too high)

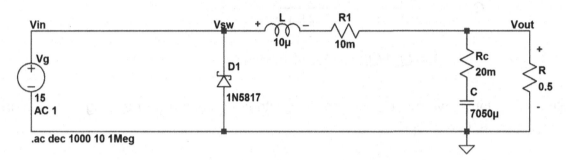

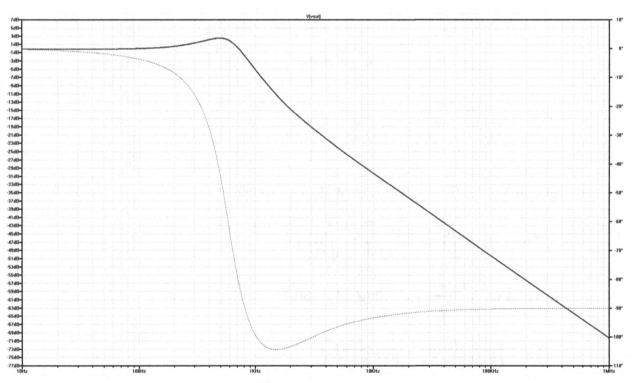

4. As we can see, it's more effective to use a De-Queing circuit, where C=2700µF is paralleled across C=470µF and R=2.5Ω (should use max R_{L_max} value to calculate Q.)

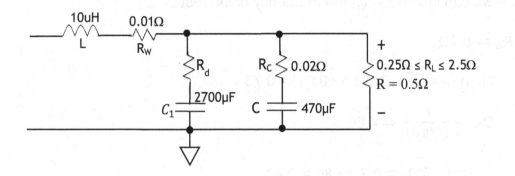

$$R_O = \sqrt{\frac{L}{C}} = \frac{1}{\sqrt{47}} \qquad \text{and} \qquad f_o = \frac{1}{2\pi\sqrt{LC}} = \frac{1}{2\pi\cdot\sqrt{10\mu H\cdot 470\mu F}} = 2.32\text{kHz}$$

$$Q_L = \frac{(R_d\|R)+R_W}{R_O} = \sqrt{47}((R_d\|2.5\Omega)+0.01)$$

$$Q_C = \frac{R_O}{(R_W\|R\|R_d)+R_C} = \frac{1}{\sqrt{47}}\cdot\frac{1}{0.01\|2.5\|R_d+0.02}$$

Note: In order to regulate the output voltage within the design specification, we need to look at D_{min} or D'_{max}, where D' cycle needs to supply enough current to maintain the regulation.

For $R_d = 0.5\Omega$

$$Q_L = \sqrt{47}\ (0.5\|2.5\Omega + 0.01) = 2.93$$

$$Q_C = \frac{1}{\sqrt{47}\ (0.01\|2.5\|0.5 + 0.02)} = 4.86$$

$$Q = Q_L\|Q_C = 2.93\|4.86 = 1.83$$

If $R_d = 0.2\Omega$

Then $Q_L = \sqrt{47}\ (0.2\|2.5\Omega + 0.01) = 1.3$

$$Q_C = \frac{1}{\sqrt{47}\ (0.03)} = 4.86$$

$$Q = Q_L\|Q_C = 1.3\|4.86 = 1.05$$

Check if $X_{C1} = \dfrac{1}{2\pi f_O\cdot C_1} < (R_d = 0.2\Omega)$ at characteristic impedance at $f_o = 2.32$kHz.

$$X_{C1} = \frac{1}{2\pi f_O\cdot C_1} = \frac{1}{2\pi\cdot 2.32\text{kHz}\cdot 2700\text{uF}} = 0.025\Omega$$

$X_{C1} = 25\text{m}\Omega$, which is $\ll 0.2\Omega$

Hence we can ignore X_{C1} @ the frequency of interest at f_o.

For $R_d = 0.1\Omega$

Then $Q_L = \sqrt{47}\,(0.1\|2.5 + 0.01) = 0.73$

$$Q_C = \frac{1}{\sqrt{47}\,(0.03)} = 4.86$$

$$\boxed{Q = Q_L\|Q_C = 0.73\|4.86 = 0.63}$$

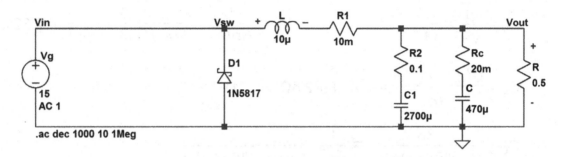

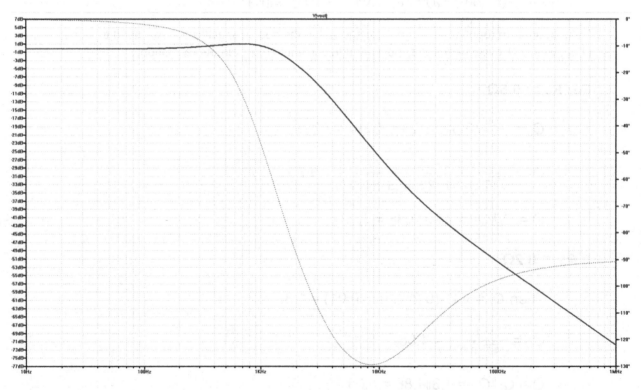

Design of A(s) compensation:

Need:

1) Determine unity bandwidth of Switch Frequency, f_{SW}, $\frac{f_S}{20} \leq f_{BW} \leq \frac{f_S}{10}$

2) Integrator

3) Zero at 1.46kHz to cancel pole @ $Q \cdot f_0 = f_{P1}$

4) Zero at 3.68kHz to cancel pole @ $\frac{f_0}{Q} = f_{P2}$

5) Pole at 16.9kHz to cancel ESR zero, f_{ESR}

6) Cancelled High frequency pole @ $\frac{1}{2} f_S$ to alternate High frequency noise (to watch phase lag at 30kHz).

When Q > 0.5, H(s) filter has two complex poles at f_0. Since 0.5 < (Q =0.63) < 1, we can set one zero before f_0 and one zero after f_0, where f_0 is calculated at 2.32kHz.

LCR Characteristic Impedance f_0 is derived as:

$$f_0 = \frac{1}{2\pi\sqrt{LC}} = \frac{1}{2\pi\sqrt{(10\mu H)(470\mu F)}} = 2.32kHz$$

For $\quad Q \cdot f_0 = 0.63(2.32kHz) = 1.46kHz = f_{P1}$ $\quad$ (to cancel it with zero at f_{z1})

$$\frac{f_0}{Q} = \frac{2.32kHz}{0.63} = 3.68kHz = f_{P2} \qquad \text{(to cancel it with zero at } f_{z2})$$

$$f_{ESR} = \frac{1}{2\pi(470uF)(0.02\Omega)} = 16.9kHz \qquad \text{(to cancel with pole at } f_{p1})$$

A(s) error amplifier design compensations are shown in Bode Plot below.

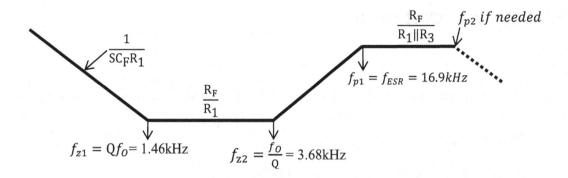

$$f_{z1} = Qf_O = 1.46\text{kHz} \qquad f_{z2} = \frac{f_O}{Q} = 3.68\text{kHz}$$

Thus, design A(s) Integrator to match 3), 4) and 5) compensation requirements:

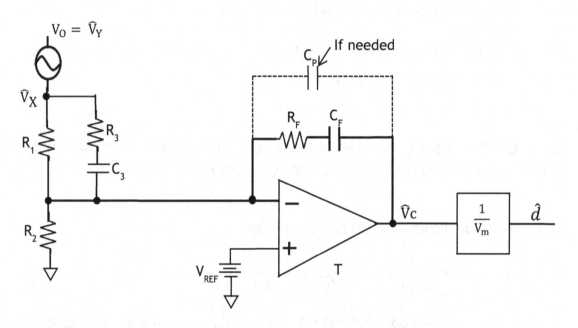

$$\frac{\widehat{V}_C}{\widehat{V}_X} \triangleq A(s) = \frac{1}{sC_FR_1} \cdot \frac{(sC_FR_F + 1)(sC_3R_1 + 1)}{(sC_3R_3 + 1)[(sC_PR_F + 1)]}$$

If needed

Calculate A(s) compensation parameters:

For f_S = 300kHz   choose $f_{BW} = \dfrac{f_S}{10}$ = 30kHz

Then we have $T_{nom} = \dfrac{V_{g\_nom}}{V_m} \cdot \dfrac{1}{SsR_1}$

$$\therefore \qquad f_{BW} = \frac{V_{g\_nom}}{V_m} \cdot \frac{1}{2\pi C_F R_1}$$

$$\therefore \quad C_F = \frac{V_{g\_nom}}{V_m} \cdot \frac{1}{2\pi \cdot f_{BW} \cdot R_1}$$

For V_O = 2.5V, then $V_O = \left(1 + \frac{R_1}{R_2}\right) \cdot V_{REF}$

$$\therefore \quad R_1 = \left(\frac{V_O}{V_{REF}} - 1\right) \cdot R_2$$

Let R_2 =10kΩ, Then $R_1 = \left(\frac{2.5}{1.25} - 1\right) 10k\Omega = 10k\Omega$

$$\boxed{R_2 = R_1 = 10k\Omega}$$

$$V_{g\_nom} = V_{in\_nom} = 15V$$

$$C_F = \frac{V_{g\_nom}}{V_m} \cdot \frac{1}{2\pi \cdot f_{BW} \cdot R_1} = \left(\frac{15}{2}\right)\left(\frac{1}{2\pi(30kHz)(10k\Omega)}\right) = 4.0 \text{ nF}$$

$$R_F = \left(\frac{1}{2\pi(Qf_o)C_F}\right) = \left(\frac{1}{2\pi(1.46kHz)(4nF)}\right) = 55 \text{ k}\Omega$$

$$C_3 = \frac{1}{2\pi(\frac{f_o}{Q})R_1} = \left(\frac{1}{2\pi(3.68kHz)(10k\Omega)}\right) = 4.3 \text{ nF}$$

$$R_3 = \frac{1}{2\pi f_{ESR}C_3} = \left(\frac{1}{2\pi(16.9kHz)(4.3nF)}\right) = 2.2 \text{ k}\Omega$$

Loop-Gain after the A(s) compensation:

Hence for Loop Gain $T_{nominal}$ we have

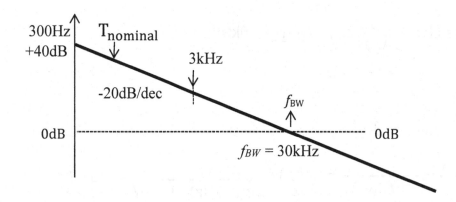

Note: For find:

(1). $F = \left.\dfrac{\hat{V}}{\hat{V}_g}\right|_{C.L} = \dfrac{GV_g}{1+T} = \dfrac{DH(s)}{1+T}$

where $GV_g \triangleq \left.\dfrac{\hat{V}}{\hat{V}_g}\right|_{O.L}$ (C.L → Closed loop, O.L → Open Loop)

(2). $\left.Z_0\right|_{C.L} = \dfrac{Z_{0\,(O.L)}}{1+T}$

14.2 Evaluating the Stability of the Buck Converter goes to DCM

What happens at light load or NO LOAD when the inductor current goes to zero before the end of the switching period, T_S, [i.e., DCM].

We must find the new Loop Gain T_{NEW} based on DCM model, where

$$T_{NEW} = \frac{\hat{V}_g}{\hat{d}} \cdot \frac{\hat{d}}{\hat{V}_C} \cdot \frac{\hat{V}_C}{\hat{V}_X}$$

Note: $\dfrac{\hat{d}}{\hat{V}_C} = \dfrac{1}{V_m}$ and $\dfrac{\hat{V}_C}{\hat{V}_X} = A(s)$ [These do not change for CCM or DCM].

The only thing that changes is $\dfrac{\hat{V}_Y}{\hat{d}}$. Set $\hat{V}_g = 0$.

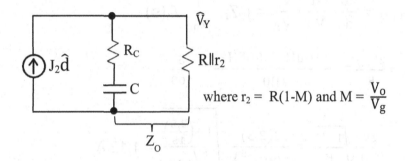

where $r_2 = R(1-M)$ and $M = \dfrac{V_o}{V_g}$

$$\boxed{\frac{\hat{V}_Y}{\hat{d}} = j_2 \cdot Z_O}$$

Finding the point that Buck Converter goes to DCM, where $D_2 = D'$.

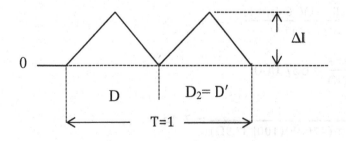

Note: $D_2 = \sqrt{K(1-M)}$

$K = \dfrac{2Lf_S}{R}$ we need to find R when $D_2 = D' = 1 - D$

But D = $\dfrac{V}{V_g}$ = M   [when D_2 = D' or CCM]

$\therefore$ D_2 = D' = $\sqrt{K(1-M)}$= $\sqrt{K \cdot D'}$

$\therefore$ D'^2= K· D'   →   K = D' = 1−D

For V_{g_nom}, then D = $\dfrac{V}{V_{g\_nom}}$ = $\dfrac{2.5V}{15V}$ = 0.167

$\therefore$ R = $\dfrac{2Lf_S}{K}$ = $\dfrac{2(10\mu H)(300kHz)}{1-0.167}$ = $\dfrac{6}{1-0.167}$ = 7.2Ω

Let's look at f_{BW} when R = 100Ω

$$T_{NEW} = \dfrac{\hat{V}_y}{\hat{d}} \cdot \dfrac{\hat{d}}{\hat{V}_C} \cdot \dfrac{\hat{V}_C}{\hat{V}_X} = j_2 Z_O \cdot \dfrac{1}{V_m} \cdot A(s)$$

$$K = \dfrac{2Lf_S}{R} = \dfrac{2(10\mu H)(300kHz)}{100} = \dfrac{6}{100}$$

$$j_2 = \dfrac{2V}{RM}\sqrt{\dfrac{1-M}{K}} = \dfrac{2(2.5)}{100\left(\frac{2.5}{15}\right)}\sqrt{\dfrac{1-\left(\frac{2.5}{15}\right)}{6/100}} = 1.12 \text{ A}$$

$$Z_O = (R\|r_2) \cdot \dfrac{(sCR_C+1)}{sC(R\|r_2 + R_C)+1} \approx (R\|r_2) \cdot \dfrac{(sCR_C+1)}{sC(R\|r_2)+1}$$

$$R\|r_2 = R\|R(1-M) = (100\|100)\left(1 - \dfrac{2.5}{15}\right) = 45.5 \text{ Ω}$$

$$T_{NEW} = \dfrac{1.12A(45.5Ω)}{2V} \cdot \dfrac{(sCR_C+1)}{sCR\|r_2+1} \cdot \dfrac{1}{sC_F R_1} \cdot \dfrac{(sC_F R_F+1)(sC_3 R_1+1)}{(sC_3 R_3+1)}$$

$$T_{NEW} = \dfrac{K_{DM}}{s} \cdot \dfrac{(sC_F R_F+1)(sC_3 R_1+1)}{(sCR\|r_2+1)}$$

$$K_{DM} = \dfrac{(1.12)(45.5)}{2(4nF)(10K)} = 637,000$$

$$f_P = \dfrac{1}{2\pi CR\|r_2} = \dfrac{1}{2\pi(470uF)(100\|83.3Ω)} = 7.44 \text{ Hz}$$

But f_{Z1} = $\dfrac{1}{(sC_F R_F + 1)}$ = 1.46kHz and f_{Z2} = $\dfrac{1}{(sC_3 R_1+1)}$ = 3.68 kHz

Assume f_{BW} > f_P , but f_{BW} < f_{Z1}

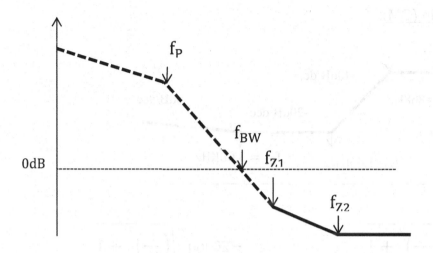

$$T_{NEW} = \frac{K_{DM}}{s} \cdot \frac{1}{(sCR\|r_2)} = 1$$

$$\therefore\ 1 = \frac{K_{DM}}{CR\|r_2} \cdot \frac{1}{(2\pi)^2 f_{BW}^2} = 1$$

$$\therefore f_{BW} = \sqrt{\frac{K_{DM}}{CR\|r_2}\frac{1}{(2\pi)^2}} = \frac{1}{2\pi}\sqrt{\frac{K_{DM}}{CR\|r_2}}$$

$$f_{BW} = \frac{1}{2\pi}\sqrt{\frac{637{,}000}{(470uF)(45.5\Omega)}} = \frac{1}{2\pi}\sqrt{\frac{637{,}000}{0.0214}}$$

$$\boxed{f_{BW} = 868\ \text{Hz}}$$

Since $f_{BW} < f_{Z1}$, we are correct in our assumption.

Let's look at $\varnothing_m$ (phase margin) at $f_{BW} = 868$Hz

$$\varnothing_m = 90^0 - \tan^{-1}\frac{868\text{Hz}}{7.44\text{Hz}} + \tan^{-1}\frac{868\text{Hz}}{1.46\text{KHz}} + \tan^{-1}\frac{868\text{Hz}}{3.68\text{KHz}}$$

$$\varnothing_m = 90^0 - 89.5^0 + 30.7^0 + 13.3^0$$

$$\boxed{\varnothing_m = 44.5^0}$$

For find Gain Margin (GM):

$$\text{GM} = -40 \log \sqrt{\left(\frac{f}{f_{BW}}\right)^2 + 1} \qquad -20 \log \sqrt{\left(\frac{f}{f_{Z1}}\right)^2 + 1}$$

$$f_{BW} < f < f_{z1} \quad \text{and} \quad f_{z1} < f < f_{z2}$$

$$\therefore \text{GM} = -40 \log \sqrt{\left(\frac{1460}{868}\right)^2 + 1} \qquad -20 \log \sqrt{\left(\frac{3.68}{1.46}\right)^2 + 1}$$

$$\boxed{\text{GM} \geq -11.66 \text{ dB} -8.66 \text{ dB} = -20.32 \text{ dB}}$$

Conclusion:

This design is stable for Continuous Conduction Mode (CCM) and Discontinuous Conduction Mode (DCM).

APPENDIX A: Common Types of Operational Amplifiers

Inverting Operational Amplifier

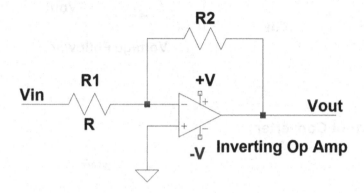

Non-inverting Operational Amplifier

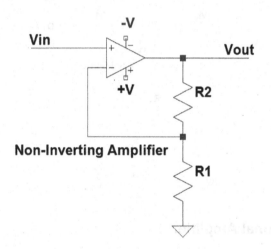

Summing Amplifier

Voltage Follower

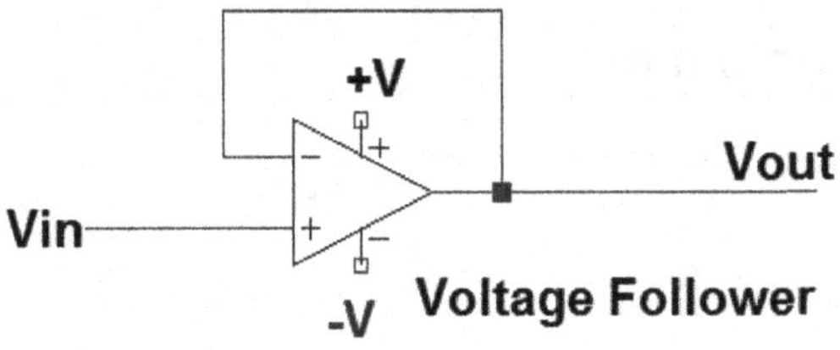

+V

Vin

Vout

-V **Voltage Follower**

Voltage to Current Converter

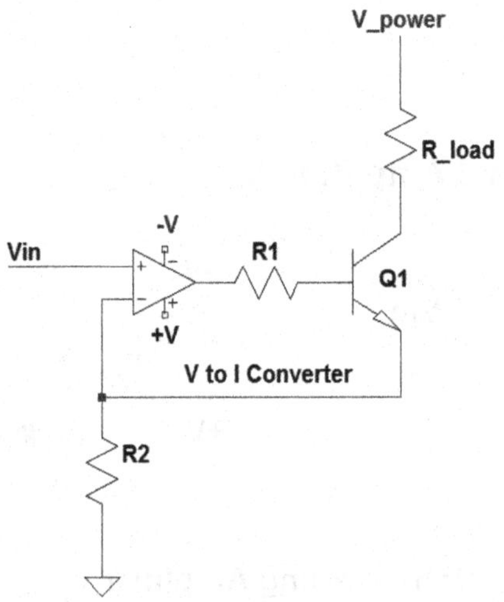

V_power

R_load

-V

Vin

R1

Q1

+V

V to I Converter

R2

Differential Operational Amplifier

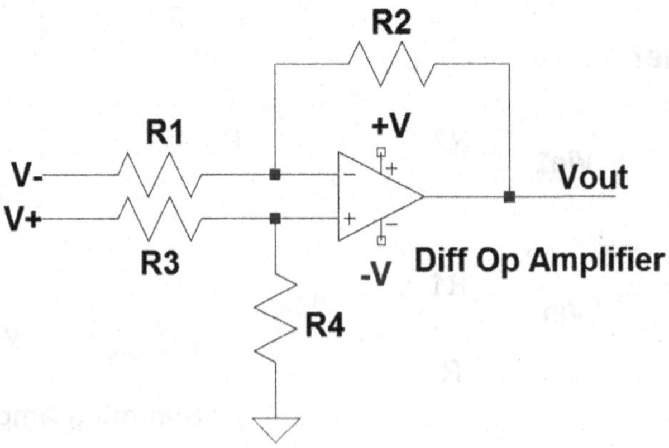

R2

R1

+V

V-

V+

Vout

R3

-V **Diff Op Amplifier**

R4

Voltage Comparator

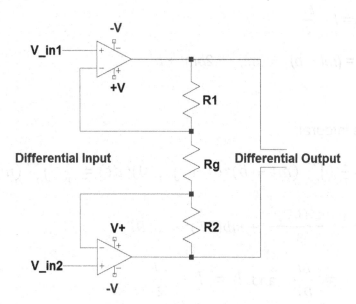

Window Comparator

Differential Input Amplifier

Appendix A-1: Where does the $\frac{1}{12}$ come from in I_{rms}^2 equation?

Where does the $\frac{1}{12}$ come from in $I_{rms}^2 = DI^2 \left[1 + \left(\frac{1}{12} \cdot \frac{\Delta I}{I}\right)^2\right]$?

Inductor's current waveform is:

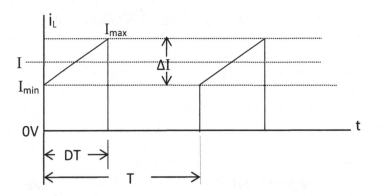

We know that $I_{rms}^2 \triangleq \frac{1}{T} \int_0^T i(t)^2 \cdot dt$

So we must find $i(t)$ in terms of I and ΔI, then find $i(t)^2$ and evaluate the integral.

From $0 \leq t \leq DT$, we find i(t) to be:

i(t) = mt + b, which is a linear equation with b as intercept.

$$m = \frac{I_{max} - I_{min}}{DT} = \frac{\Delta I}{DT} \qquad \text{and} \qquad b = I_{min}$$

But $I_{min} = I - \dfrac{\Delta I}{2}$

$$\therefore\ i^2(t) = (mt + b)^2 = (mt)^2 + 2bmt + b^2$$

Evaluating the integral:

$$I_{rms}^2 = \frac{1}{T} \left(\int_0^{DT} (mt + b)^2 dt + \int_{DT}^T (0)^2 dt\right) = \frac{1}{T} \left(\int_0^{DT} (m^2 t^2 + 2mbt + b^2) dt\right)$$

$$I_{rms}^2 = \frac{1}{T} \left(\frac{m^2 (DT)^3}{3} + mb(DT)^2 + b^2 DT\right)$$

Substituting: $m = \dfrac{\Delta I}{DT}$ and $b = I - \dfrac{\Delta I}{2}$

$$I_{rms}^2 = \frac{1}{T} \left(\left(\frac{\Delta I}{DT}\right)^2 \cdot \frac{(DT)^3}{3} + \left(\frac{\Delta I}{DT}\right)\left(I - \frac{\Delta I}{2}\right)(DT)^2 + \left(I - \frac{\Delta I}{2}\right)^2 DT \right)$$

$$I_{rms}^2 = \frac{1}{T} \left[\left(\frac{\Delta I^2}{3}(DT) + \Delta I(I)(DT) - \frac{\Delta I^2}{2} DT + (I^2 - \Delta I(I) + \frac{\Delta I^2}{4})DT \right] \right.$$

Combine common terms:

$$I_{rms}^2 = D \left(\frac{\Delta I^2}{3} - \frac{\Delta I^2}{2} + \frac{\Delta I^2}{4} + I^2 \right)$$

$$I_{rms}^2 = D\left(\Delta I^2\left[\frac{1}{3} - \frac{1}{2} + \frac{1}{4}\right] + I^2\right)$$

Common denominator is 12:

$$\therefore \quad I_{rms}^2 = D \left(\Delta I^2\left(\frac{4}{12} - \frac{6}{12} + \frac{3}{12}\right) + I^2 \right)$$

$$I_{rms}^2 = D \left(\frac{\Delta I^2}{12} + I^2 \right)$$

Factor out I^2 and we have:

$$\boxed{I_{rms}^2 = DI^2 \left[1 + \frac{1}{12} \cdot \left(\frac{\Delta I}{I}\right)^2 \right]}$$

APPENDIX A-2: Superposition Methods

To ascertain contribution of each individual source, we need to take a look at output responses with respect to each individual source. At the same time, we must set remaining sources to zero, where evaluated variables are independent from the rest of circuits.

1. **For Voltage Source:** Short "unwanted dependent" voltage sources, since the internal impedance of ideal voltage source is zero.
2. **For Current Source:** Open "unwanted dependent" sources, since the internal impedance of ideal current source is infinite.
3. Evaluate voltage, current or transfer function of each individual source.
4. Overall output or system responses are sum of individual responses from superposition methods.

(Note: Superposition only works for voltage and current. It does not apply to power because Power terms are not linear circuit. The superposition theorem is very important in circuit analysis. It is used in converting any circuit into its Thevenin Equivalent or Thevenin current or Thervenin Impedance.

Example 1: Determine current flow and magnitude via R2 = ?

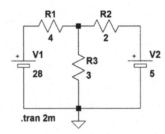

Step 1: Short V_1, redraw the circuit. Determine current flow via R_2.

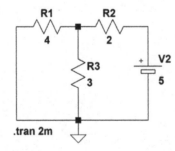

$I_{R2} = V_2/(R_2 + R_1 \| R_3) = 1.346$ A$_{dc}$, and current flow is from right to left.

Step 2: Restore the circuit, short V_2, redraw the circuit, and determine the current flow through R_2.

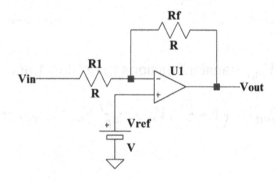

$I_{R1} = V_1/(R_1 + R_2 \| R_3) = 5.38 \text{ A}_{dc}$

$I_{R2} = (R_3/(R_2 + R_3))^* I_{R1} = 3.228 \text{ A}_{dc}$, and current flow is from left to right.

Step 3: Determine the overall R_2's current magnitude and direction

$I_{overall\_R2} = 3.228 - 1.346 = 1.882 \text{ A}_{dc}$, and current flow is from left to right.

Example 2: Calculate Transfer function of inverter Op Amplifier.

i.   Derived from conventional method:

$$\frac{V_{out} - V_-}{R_f} = \frac{V_- - V_{in}}{R_1} \quad \rightarrow \quad R_1 V_{out} - R_1 V_- = R_f V_- - R_f V_{in}$$

$$R_1 V_{out} = (R_f + R_1)V_- - R_f V_{in} \quad \rightarrow \quad V_{out} = \frac{(R_1 + R_f)}{R_1} \cdot V_- - \frac{R_f}{R_1} \cdot V_{in}$$

Since $V_- = V_{ref}$, then $V_{out} = (1 + \frac{R_f}{R_1}) \cdot V_{ref} - \frac{R_f}{R_1} V_{in}$

ii. Derived from the Superposition method:

Step 1: Short V_{ref} to GND, and redraw the circuit (inverter op-amp).

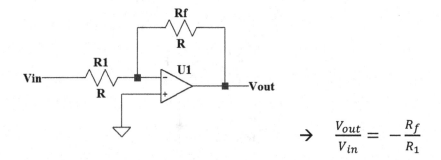

$$\rightarrow \quad \frac{V_{out}}{V_{in}} = -\frac{R_f}{R_1}$$

Step 2: Restore the circuit, short V_{in} to GND, and redraw the circuit

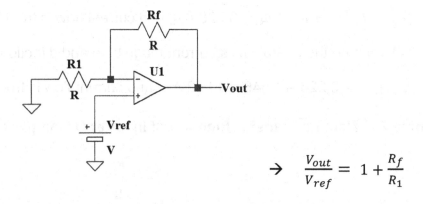

$$\rightarrow \quad \frac{V_{out}}{V_{ref}} = 1 + \frac{R_f}{R_1}$$

Step 3: Add V_{out} transfer functions from Step 1 and 2 above.

$$V_{out} = \left(1 + \frac{R_f}{R_1}\right) V_{ref} - \frac{R_f}{R_1} V_{in} = V_{ref} + \frac{R_f}{R_1}\left(V_{ref} - V_{in}\right)$$

APPENDIX A-3: Modular Gain Derivation

NEXT WE MUST FIND THE MODULATOR GAIN: $M(s) = \dfrac{\hat{d}}{\hat{v}_c}$

Our first approach is to use a ramp applied to comparator.

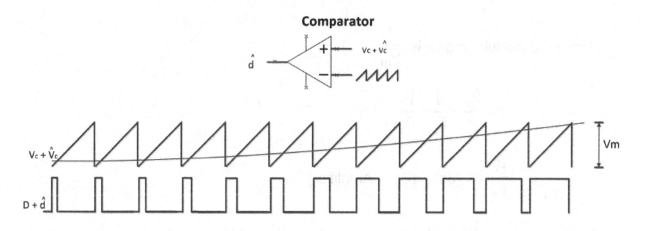

as we can see V_m is the range of the control signal V_c, required to sweep to duty ratio D, from 0 to -1.

We note: For $\dfrac{V_c}{V_m} = D$ or $\dfrac{1}{V_m} = \dfrac{1}{V_c}$

V_c	D
$V_c = 1/4\ V_g$	D = 1/4
$V_c = 1/2\ V_g$	D = 1/2
$V_c = 3/4\ V_g$	D = 3/4
$V_c = 1\ V_g$	D = 1

If we now perturbate the control signal V_c, then $V_c = V_c + \hat{v}_c$ and $d = D + \hat{d}$

But V_m = constant amplitude

Hence: $\dfrac{D + \hat{d}}{V_c + \hat{V}_c} = \dfrac{1}{V_m}$

DC Term: $\dfrac{D}{V_c} = \dfrac{1}{V_m}$ (set AC Terms to zero)

AC Term: $\dfrac{\hat{d}}{\hat{V}_c} = \dfrac{1}{V_m}$ (set DC Terms to zero)

Hence the modulator gain is $\dfrac{1}{V_m}$

$$T = \frac{\hat{V}_y}{\hat{d}} \cdot \frac{\hat{d}}{\hat{V}_c} \cdot \frac{\hat{V}_c}{\hat{V}_x}$$

$$\frac{\hat{V}_c}{\hat{V}_x} = A(s) \text{(Error Amplifier)}$$

$$T = \frac{V_g}{V_m} \cdot \frac{N_s}{N_p} \cdot H(s) \cdot A(s)$$

We desire a single pole roll off and parameters to control the loop gain at low frequency.

$$\therefore A(s) = \frac{K}{s} \cdot H(s)$$

Then $\quad T = \dfrac{V_g}{V_m} \cdot \dfrac{N_s}{N_p} \cdot H(s) \cdot A(s) = \dfrac{V_g}{V_m} \cdot \dfrac{N_s}{N_p} \cdot H(s) \cdot \dfrac{K}{s} \cdot H(s)$

$$\boxed{T = \frac{V_g}{V_m} \cdot \frac{N_s}{N_p} \cdot \frac{K}{s} = \frac{K'}{s}} \text{where} \quad \frac{K'}{s} = \frac{V_g}{V_m} \cdot \frac{N_s}{N_p} \cdot K$$

For cross over frequency at unity gain, where T_{dB} = 0dB, which T=1.

$$1 = \frac{K'}{s} = \frac{K'}{2\pi f_c} \quad \rightarrow \quad f_c = \frac{K'}{2\pi}$$

APPENDIX A-4: Poles and Zeros Passive Filter Transfer Function

1. Resistor Voltage Divider:

$$\rightarrow \quad \frac{V_{out}}{V_{in}} = R_2 \| R_1 = \frac{R_2}{R_1 + R_2}$$

Transfer Function: **resistor voltage divider**

Phase Angle: $\tan\theta = 1 \quad \rightarrow \quad \theta = \tan^{-1}(1) = 45°$

2. RC Low Pass Filter:

$$\rightarrow \quad \frac{V_{out}}{V_{in}} = X_c \| R = \frac{1}{sCR + 1} = \frac{1}{\frac{s}{\omega_p} + 1}$$

Transfer Function:

$$\frac{V_{out}}{V_{in}} R \| X_c \qquad \rightarrow \qquad \frac{V_{out}}{V_{in}} = \frac{X_c}{X_c + R} \qquad \rightarrow \qquad \frac{V_{out}}{V_{in}} = \frac{1}{1 + \frac{R}{X_c}}$$

$$X_c = \frac{1}{sC} \qquad \rightarrow \qquad \frac{R}{X_c} = sRC \qquad \rightarrow \qquad \frac{V_{out}}{V_{in}} = \frac{1}{1 + sRC}$$

$$s = j\omega = j2\pi f \qquad \rightarrow \qquad f_p = \frac{1}{2\pi RC} \qquad \rightarrow \qquad \omega_p = \frac{1}{RC}$$

$$\frac{V_{out}}{V_{in}} = \frac{1}{1+sRC} = \frac{1}{1+\frac{s}{\omega_p}} = \frac{1}{1+\frac{\omega}{\omega_p}} = \frac{1}{1+\frac{j2\pi f}{j2\pi f_p}} = \frac{1}{\frac{f}{f_p}+1}$$

Phase Angle:

$$\frac{V_{out}}{V_{in}} = \frac{1}{1+j\omega RC} \qquad \rightarrow \qquad \tan\theta = \frac{|j\omega RC|}{1} = \omega RC = \frac{\omega}{\omega_p} = \frac{f}{f_p}$$

$$\tan\theta = \frac{\omega}{\omega_p} = \frac{f}{f_p} \qquad \textbf{or} \qquad \theta = \tan^{-1}\left(\frac{\omega}{\omega_p}\right) = \tan^{-1}\left(\frac{f}{f_p}\right)$$

3. RC High Pass Filter:

$$\rightarrow \qquad \frac{V_{out}}{V_{in}} = R\|X_c = \frac{sCR}{sCR+1}$$

Transfer Function:

$$\frac{V_{out}}{V_{in}} = R\|X_c \quad \rightarrow \quad \frac{V_{out}}{V_{in}} = \frac{R}{X_c + R} \quad \rightarrow \quad \frac{V_{out}}{V_{in}} = \frac{1}{1 + \frac{X_c}{R}}$$

$$1 + \frac{X_c}{R} = 1 + \frac{1}{sCR} = \frac{1 + sCR}{sCR} \quad \rightarrow \quad \frac{V_{out}}{V_{in}} = \frac{sCR}{1 + sCR}$$

$$f_p = \frac{1}{2\pi RC} \quad \rightarrow \quad \omega_p = \frac{1}{RC}$$

$$\frac{V_{out}}{V_{in}} = \frac{\frac{s}{\omega_p}}{1 + \frac{s}{\omega_p}} = \frac{\frac{s}{\omega_p}}{\frac{\omega_p + s}{\omega_p}} = \frac{s}{\omega_p + s} = \frac{1}{\frac{s}{\omega_p} + 1} = \frac{1}{\frac{\omega}{\omega_p} + 1} = \frac{1}{\frac{f}{f_p} + 1}$$

Phase Angle:

$$\frac{V_{out}}{V_{in}} = \frac{sCR}{1 + sCR} = \frac{j\omega RC}{1 + j\omega RC}$$

$$\frac{V_{out}}{V_{in}} = \frac{j\omega RC + \omega RC}{1 + (\omega RC)^2} = \frac{j\omega RC}{1 + (\omega RC)^2} + \frac{\omega RC}{1 + (\omega RC)^2}$$

$$\tan \theta = \frac{\left|\frac{j\omega RC}{1 + (\omega RC)^2}\right|}{\frac{\omega RC}{1 + (\omega RC)^2}} = \frac{|j\omega RC|}{\omega RC} = 1$$

$$\theta = \tan^{-1}(1) = 45°$$

$$\rightarrow \quad \frac{V_{out}}{V_{in}} = R_2 \| (X_{c1} + R_1) = \frac{sC_1R_2}{sC_1(R_2 + R_2) + 1}$$

Transfer Function:

$$\frac{V_{out}}{V_{in}} = R_2 \| (X_{c1} + R_1) \quad \rightarrow \quad \frac{V_{out}}{V_{in}} = \frac{R_2}{X_{c1} + R_1 + R_2} = \frac{1}{\frac{X_{c1}}{R_2} + \frac{R_1}{R_2} + 1}$$

$$\frac{X_{c1}}{R_2} + \frac{R_1}{R_2} = \frac{1}{sC_1R_2} + \frac{R_1}{R_2} = \frac{1 + sC_1R_1}{sC_1R_2}$$

$$\frac{X_{c1}}{R_2} + \frac{R_1}{R_2} + 1 = \mathbf{1} + \frac{1 + sC_1R_1}{sC_1R_2} = \frac{1 + sC_1R_1 + sC_1R_2}{sC_1R_2} = \frac{1 + sC_1(R_1 + R_2)}{sC_1R_2}$$

$$\frac{V_{out}}{V_{in}} = \frac{1}{\frac{X_{c1}}{R_2} + \frac{R_1}{R_2} + 1} = \frac{sC_1R_2}{1 + sC_1(R_1 + R_2)}$$

Phase Angle:

$$\frac{V_{out}}{V_{in}} = \frac{sC_1R_2}{1 + sC_1(R_1 + R_2)} = \frac{j\omega C_1R_2}{1 + j\omega C_1(R_1 + R_2)} = \frac{j\omega C_1R_2 \left(1 - j\omega C_1(R_1 + R_2)\right)}{1 + (\omega C_1(R_1 + R_2))^2}$$

$$\frac{V_{out}}{V_{in}} = \frac{j\omega C_1R_2}{1 + (\omega C_1(R_1 + R_2))^2} - \frac{j\omega C_1R_2 \left(j\omega C_1(R_1 + R_2)\right)}{1 + (\omega C_1(R_1 + R_2))^2}$$

$$\frac{V_{out}}{V_{in}} = \frac{j\omega C_1 R_2}{1 + (\omega C_1(R_1 + R_2))^2} + \frac{\omega C_1 R_2 \, (\omega C_1(R_1 + R_2))}{1 + (\omega C_1(R_1 + R_2))^2}$$

$$\tan\theta = \frac{\left|\frac{j\omega C_1 R_2}{1 + (\omega C_1(R_1 + R_2))^2}\right|}{\frac{\omega C_1 R_2 \, (\omega C_1(R_1 + R_2))}{1 + (\omega C_1(R_1 + R_2))^2}} = \frac{\omega C_1 R_2}{\omega C_1 R_2 \, (\omega C_1(R_1 + R_2))} = \frac{1}{\omega C_1(R_1 + R_2)}$$

$$\theta = \tan^{-1}\left(\frac{1}{\omega C_1(R_1 + R_2)}\right)$$

$$\rightarrow \quad \frac{V_{out}}{V_{in}} = (X_{c2} + R_2) \| (X_{c1} + R_1) = \frac{X_{c2} + R_2}{(X_{c2} + R_2) + (X_{c1} + R_1)}$$

Transfer Function:

$$\frac{V_{out}}{V_{in}} = (X_{c2} + R_2) \middle\| (X_{c1} + R_1) = \frac{X_{c2} + R_2}{(X_{c2} + R_2) + (X_{c1} + R_1)} = \frac{1}{1 + \frac{(X_{c1} + R_1)}{(X_{c2} + R_2)}}$$

$$\frac{(X_{c1} + R_1)}{(X_{c2} + R_2)} = \frac{\frac{sC_1 R_1 + 1}{sC_1}}{\frac{sC_2 R_2 + 1}{sC_2}} = \frac{C_2(sC_1 R_1 + 1)}{C_1(sC_2 R_2 + 1)} = \frac{sC_1 C_2 R_1 + C_2}{sC_1 C_2 R_2 + C_1}$$

$$1 + \frac{(X_{c1} + R_1)}{(X_{c2} + R_2)} = 1 + \frac{sC_1 C_2 R_1 + C_2}{sC_1 C_2 R_2 + C_1} = \frac{sC_1 C_2 R_2 + C_1 + sC_1 C_2 R_1 + C_2}{sC_1 C_2 R_2 + C_1}$$

$$\frac{1}{1 + \frac{(X_{c1} + R_1)}{(X_{c2} + R_2)}} = \frac{sC_1 C_2 R_2 + C_1}{sC_1 C_2 R_2 + C_1 + sC_1 C_2 R_1 + C_2} = \frac{C_1(sC_2 R_2 +)}{C_1 + C_2 + sC_1 C_2(R_2 + R_1)}$$

$$\frac{1}{1 + \frac{(X_{c1}+R_1)}{(X_{c2}+R_2)}} = \frac{C_1}{C_1+C_2} \cdot \frac{(sC_2R_2+1)}{1 + \frac{sC_1C_2(R_2+R_1)}{C_1+C_2}} = \frac{C_1}{C_1+C_2} \cdot \frac{(sC_2R_2+1)}{1 + s(C_1\|C_2)(R_2+R_1)}$$

$$\frac{V_{out}}{V_{in}} = (X_{c2} + R_2)\|(X_{c1} + R_1) = \frac{C_1}{C_1+C_2} \cdot \frac{(sC_2R_2+1)}{1 + s(C_1\|C_2)(R_2+R_1)}$$